LASER RAMAN SPECTROMETRY

ANALYTICAL APPLICATIONS

ELLIS HORWOOD SERIES IN ANALYTICAL CHEMISTRY

Series Editors: Dr. R. A. CHALMERS and Dr. MARY MASSON, University of Aberdeen

Consultant Editor: Prof. J. N. MILLER, Lougborough University of Technology

S. Allenmark — **Chromatographic Enantioseparation — Methods and Applications**
G. E. Baiulescu & V. V. Cosofret — **Application of Ion Selective Membrane Electrodes in Organic Analysis**
G. E. Baiulescu, P. Dumitrescu & P. Gh. Zugravescu — **Sampling**
G. E. Baiulescu, C. Patroescu & R. A. Chalmers — **Education and Teaching in Analytical Chemistry**
G. I. Bekov & V. S. Letokhov — **Laser Resonant Photoionization Spectroscopy for Trace Analysis**
S. Bance — **Handbook of Practical Organic Microanalysis**
H. Baranska, A. Łabudzińska & J. Terpiński — **Laser Raman Spectrometry**
K. Beyermann — **Organic Trace Analysis**
O. Budevsky — **Foundations of Chemical Analysis**
J. Buffle — **Complexation Reactions in Aquatic System: An Analytical Approach**
D. T. Burns, A. Townshend & A. G. Catchpole — **Inorganic Reaction Chemistry Volume 1: Systematic Chemical Separation**
D. T. Burns, A. Townshend & A. H. Carter — **Inorganic Reaction Chemistry: Reactions of the Elements and their Compounds: Volume 2, Part A: Alkali Metals to Nitrogen and Volume 2, Part B: Osmium to Zirconium**
S. Caroli — **Improved Hollow Cathode Lamps for Atomic Spectroscopy**
J. Churáček — **New Trends in the Theory & Instrumentation of Selected Analytical Methods**
R. Czoch & A. Francik — **Instrumental Effects in Homodyne Electron Paramagnetic Resonance Spectrometers**
T. E. Edmonds — **Interfacing Analytical Instrumentation with Microcomputers**
J. K. Foreman & P. B. Stockwell — **Automatic Chemical Analysis**
Z. Galus — **Fundamentals of Electrochemical Analysis**
J. Gasparič & J. Churáček — **Laboratory Handbook of Paper and Thin Layer Chromatography**
S. Görög — **Steroid Analysis in the Pharmaceutical Industry**
T. S. Harrison — **Handbook of Analytical Control of Iron and Steel Production**
J. P. Hart — **Electroanalysis of Biologically Important Compounds**
T. F. Hartley — **Computerized Quality Control: Programs for the Analytical Laboratory**
Saad S. M. Hassan — **Organic Analysis using Atomic Absorption Spectrometry**
M. H. Ho — **Analytical Methods in Forensic Chemistry**
Z. Holzbecher, L. Diviš, M. Král, L. Šůcha & F. Vláčil — **Handbook of Organic Reagents in Inorganic Chemistry**
A. Hulanicki — **Reactions of Acids and Bases in Analytical Chemistry**
David Huskins — **General Handbook of On-line Process Analysers**
David Huskins — **Quality Measuring Instruments in On-line Process Analysis**
J. Inczédy — **Analytical Applications of Complex Equilibria**
Z. K. Jelínek — **Particle Size Analysis**
M. Kaljurand & E. Küllik — **Computerized Multiple Input Chromatography**
R. Kalvoda — **Operational Amplifiers in Chemical Instrumentation**
I. Kerese — **Methods of Protein Analysis**
S. Kotrlý & L. Šůcha — **Handbook of Chemical Equilibria in Analytical Chemistry**
J. Kragten — **Atlas of Metal-ligand Equilibria in Aqueous Solution**
A. M. Krstulović — **Quantitative Analysis of Catecholamines and Related Compounds**
F. J. Krug & E. A. G. Zagatto — **Flow Injection Analysis in Agriculture & Environmental Science**
V. Linek & V. Vacek — **Measurement of Oxygen by Membrane-Covered Probes**
C. Liteanu & S. Gocan — **Gradient Liquid Chromatography**
C. Liteanu & I. Rîcă — **Statistical Theory and Methodology of Trace Analysis**
Z. Marczenko — **Separation and Spectrophotometric Determination of Elements**
M. Meloun, J. Havel & E. Högfeldt — **Computational Methods in Potentiometry and Spectrophotometry**
O. Mikeš — **Laboratory Handbook of Chromatographic and Allied Methods**
J. C. Miller & J. N. Miller — **Statistic for Analytical Chemistry**
J. N. Miller — **Fluorescence Spectroscopy**
J. N. Miller — **Modern Analytical Chemistry**
J. Minczewski, J. Chwastowska & R. Dybczyński — **Separation and Preconcentration Methods in Inorganic Trace Analysis**

(*continued on p.* 271)

LASER RAMAN SPECTROMETRY
ANALYTICAL APPLICATIONS

Halina BARAŃSKA
Anna ŁABUDZIŃSKA
and
Jacek TERPIŃSKI

Translation Editor:
J. R. Majer

ELLIS HORWOOD LIMITED
Publishers · Chichester

Halsted Press: a division of
JOHN WILEY & SONS
New York · Chichester · Brisbane · Toronto

PWN—POLISH SCIENTIFIC PUBLISHERS
Warsaw

English edition first published in 1987 in coedition between
ELLIS HORWOOD LIMITED
Market Cross House, Cooper Street, Chichester, West Sussex, PO19 IEB; England

and

PWN—POLISH SCIENTIFIC PUBLISHERS
Warsaw, Poland

The Horwood publisher's colophon is reproduced from James Gillison's drawing of the ancient Market Cross, Chichester

Translated by *Jerzy Lipski* (Chapters 3, 4, 5, 7 and 9), *Krzysztof Radziwiłł* (Chapters 1, 2 and 6) and *Piotr Dryjański* (Chapter 8) from the Polish edition *Laserowa spektrometria ramanowska, Zastosowania analityczne* PWN—Państwowe Wydawnictwo Naukowe, Warszawa 1981

Distributors:

Australia, New Zealand, South-east Asia:
Jacaranda-Wiley Ltd., Jacaranda Press,
JOHN WILEY & SONS INC.,
G.P.O. Box 859, Brisbane, Queensland 4001, Australia

Canada:
JOHN WILEY & SONS CANADA LIMITED
22 Worcester Road, Rexdale, Ontario, Canada

Europe, Africa:
JOHN WILEY & SONS LIMITED
Baffins Lane, Chichester, West Sussex, England

Albania, Bulgaria, Cuba, Czechoslovakia, German Democratic Republic, Hungary, Korean People's Democratic Republic, Mongolia, People's Republic of China, Poland, Romania, the U.S.S.R., Vietnam and Yugoslavia:
ARS POLONA—Foreign Trade Enterprise
Krakowskie Przedmieście 7, 00-068 Warszawa, Poland

North and South America and the rest of the world:
Halsted Press: a division of
JOHN WILEY & SONS
605 Third Avenue, New York, N.Y. 10016, U.S.A.

British Library Cataloguing in Publication Data

Barańska, Halina
Laser raman spectrometry: analytical applications.
— (Ellis Horwood series in analytical chemistry).
1. Raman spectroscopy
2. Laser spectroscopy
3. Chemistry, analytic
I. Title II. Labudzińska, Anna III. Terpiński, Jacek
IV. Majer, J. R. V. Laserowa spektrometria ramanowska

English

543'.08584 QD96.R34
ISBN 0-85312-339-X (Ellis Horwood Limited)
ISBN 0-470-20829-5 (Halsted Press)

Library of Congress Card. No. 87-2892

Printed in Poland

Table of Contents

1

An introduction to Raman scattering

H. Barańska

1.1 HISTORICAL OUTLINE

The early nineteen-twenties brought a rapid development of the quantum-mechanical theory of light scattering. The Austrian physicist Smekal [1] was the first of a range of outstanding physicists which included Kramers and Heisenberg [2], Schrödinger [3] and Dirac [4], to predict in 1923 that radiation scattered from molecules contains not only photons with the incident-photon frequency but also some with a changed frequency. This prediction found experimental confirmation in 1928, when physicists from several countries made tests with samples in the three states of matter: liquid, solid and gas. The first positive results of experiments with liquid benzene were obtained by Chandrasekhara Venkata Raman—an Indian physicist [5]. Raman was awarded the Nobel prize and the phenomenon he discovered now carries his name. The Soviet physicists Landsberg and Mandelstam [6] discovered a new kind of scattering in quartz crystals, but the French physicists Rocard [7] and Cabannes [8] were less lucky and observed no such phenomenon in their experiments with gases.

The instruments used for recording Raman spectra consisted for many years of the units presented schematically in Fig. 1.1. The radiation from a mercury lamp (*1*) passed through a filter (*2*), consisting, for example, of aqueous sodium nitrite solution. The monochromatic light obtained in this way was used for illuminating the sample (*3*). The scattered radiation was observed at an angle of 90° to the incident beam. It was dispersed by a glass prism (*4*) and registered on a photographic plate (*5*). The result-

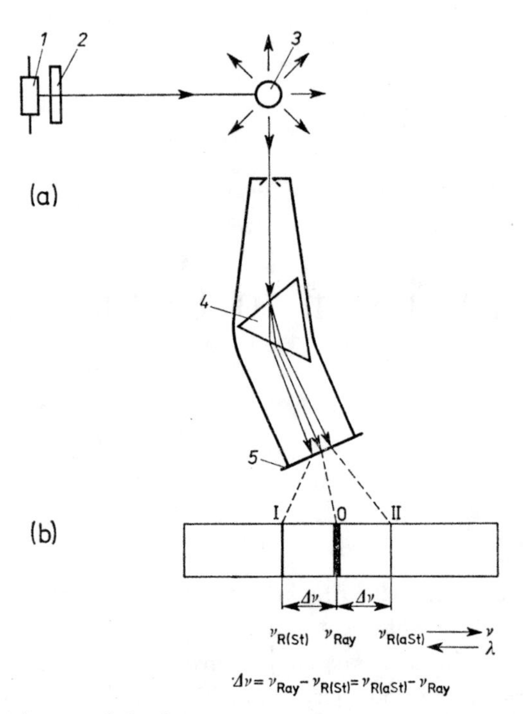

Fig. 1.1—(a) Diagram of the first Raman spectrographs. (*1*) Mercury-arc lamp. (*2*) Solution filter (e.g. saturated sodium nitrite solution transmitting radiation corresponding to the mercury line at 435,8 nm). (*3*) Sample. (*4*) Prism. (*5*) Photographic plate. (b) Raman spectrum registered on photographic plate. (0) Rayleigh scattering line of ν_{Ray} equal to the exciting line frequency, (I) Raman Stokes scattering band of $\nu_{R(St)}$, (II) Raman anti-Stokes scattering band of $\nu_{R(aSt)}$.

ing spectrum consisted of a very strong line corresponding to the incident radiation wavelength (Rayleigh scattering) and of weak and very weak Raman scattering bands* distributed symmetrically on both sides of that line. *The frequency shift of those bands with respect to the Rayleigh line*

* The band corresponding to the frequency $\nu_{R(St)}$ is referred to as the Stokes band because of its shift from the exciting line towards longer wavelengths. Such a shift is observed in fluorescence and is called the Stokes shift after the first investigator of fluorescence. The Raman scattering band corresponding to the frequency $\nu_{R(aSt)}$ and known as anti-Stokes scattering, is shifted from the exciting band towards shorter wavelengths.

is characteristic and constant for the given substance and is independent of the incident band frequency.

Since then Raman spectroscopy has had many ups and downs. It has contributed to the development of our knowledge of rotational–vibrational spectra, and has been an important analytical method for liquid hydrocarbons and aqueous solutions. Although in chemical analysis it has been outstripped by infrared absorption and gas chromatography, it has been reborn as laser Raman spectrometry since the discovery of lasers as ideal sources of monochromatic radiation [9]. Today, laser Raman spectrometry is becoming an important and indispensable method providing information about molecules and their changes in chemical processes. Laser spectrometers allow the recording of Raman spectra for very small samples, of the order of a few milligrams or a few hundred microlitres, over a wide range of temperatures and pressures. Liquids, solutions, crystalline powders and single crystals can all be examined. Coloured samples can also be investigated and Raman scattering from a sample surface recorded.

Raman spectra allow the identification of molecules and a study of their structure, the characterization of chemical reactions, and the determination of some thermodynamic functions as well as qualitative and quantitative analysis of molecular systems.

The properties of the laser beam have not only simplified the recording of Raman spectra but have also made it possible to demonstrate a range of effects unknown earlier, such as the resonance Raman effect (cf. Section 1.8), which open new analytical possibilities.

1.2 INTERPRETATION OF RAMAN SCATTERING IN TERMS OF THE QUANTUM THEORY

Raman scattering is one of the phenomena that occur when electromagnetic radiation interacts with a molecule. Electromagnetic radiation exhibits the properties of both a stream of energetic particles and of a wave. Thus Raman scattering can also be described in two ways. The historically earlier interpretation is based on the wave theory, which was the classic approach to electromagnetic radiation. In the present book this will be given, however, only after the quantum interpretation, which has a more immediate appeal to the imagination of the present-day reader.

Energy particles are generally known as quanta, and in the case of electromagnetic radiation also as photons. The relationship between the energy E of the photon and its frequency ν is given by the Planck formula:

$$E = h\nu$$

where h is the Planck constant.

Interaction of the photon with a molecule can yield three phenomena: absorption, emission and scattering.

Absorption takes place if the photon energy corresponds to the difference between two stationary energy levels of the molecule.

Emission occurs from the excited molecule, after a time equal to the lifetime of the molecule in the initial excited level or a longer time (e.g. in the case of a transition to a metastable level followed by phosphorescence). The energy of the photon emitted also corresponds to the difference between two stationary energy levels of the molecule.

Scattering of radiation takes place as an "immediate" (within about 10^{-14} s) effect of interaction of the photon and molecule when the photon energy does not correspond to the difference between any two stationary energy levels of the molecule. Scattering can proceed without change in the energy of the incident photon (*Rayleigh scattering*) or with change of that energy (*Raman scattering*).

The normal Raman effect, which is the main object of interest in this book, is produced as a result of interaction of a molecule and a photon of visible light with an energy distinctly lower than the energy difference between the first excited and ground electronic levels in the molecule.

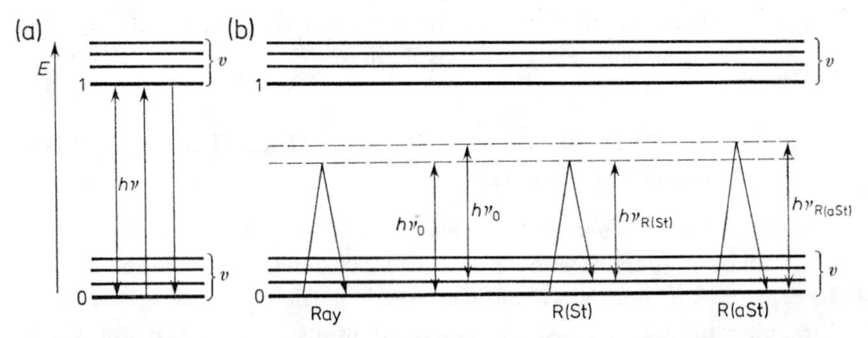

Fig. 1.2—(a) Resonance fluorescence. (b) (Ray) — Rayleigh, (R) — Raman (St- and aSt-) scattering. (0) Ground electronic state. (1) First excited electronic state. (v) Vibrational levels.

In Fig. 1.2 the difference between the phenomena discussed is illustrated by comparing resonance fluorescence and scattering.

Resonance fluorescence consists of two single-photon phenomena which can be independently observed: absorption of the photon incident on the molecule, and emission of an identical photon. The two phenomena are separated in time by a period corresponding to the lifetime (of the order of 10^{-8} s) of the molecule in the excited state.

Scattering is a two-photon process [10, 11] which cannot be experimentally separated into two single-photon steps of absorption and emission. The energy levels shown in the figure by a dashed line are non-stationary (virtual) levels of the molecule. The photon of initial energy $h\nu_0$ interacting with the molecule may proceed with unchanged energy (Rayleigh scattering), or have decreased energy $h\nu_{R(St)}$ (Raman Stokes scattering) or increased energy $h\nu_{R(aSt)}$ (Raman anti-Stokes scattering).

In the description of light-scattering phenomena the energy changes of the molecule are often denoted by vertical lines between the stationary and non-stationary levels (Figs. 1.3 and 1.4) in the same way as for transitions between stationary levels in the phenomena of absorption and emission (see Fig. 1.2, resonance fluorescence). In Fig. 1.2 the transitions for the scattering phenonenon are marked by non-vertical lines to underline its two-photon character. To understand the whole mechanism of

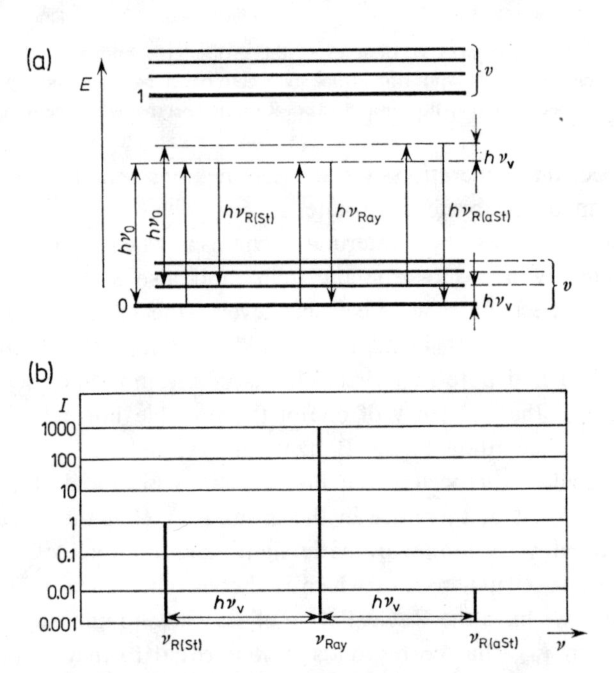

Fig. 1.3—(a) Diagram of scattering during illumination of the sample with monochromatic light. (b) Part of the resulting spectrum.

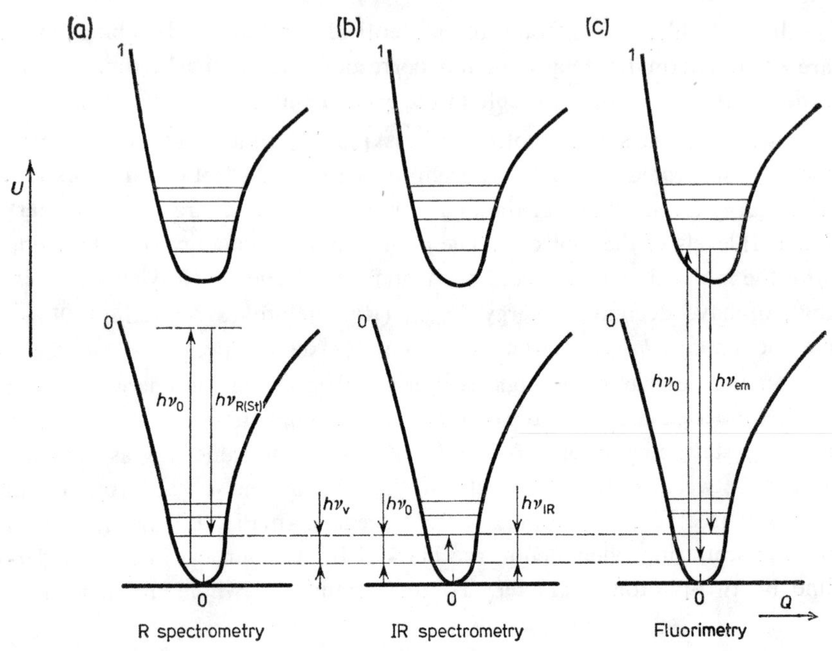

Fig. 1.4—Potential energy (U) curves of the ground (0) and excited (1) states with the energy levels and transitions between them as observed by infrared spectrometry, fluorimetry and Raman spectrometry methods.

Raman spectrum generation, we must remember that the bond in the molecule can both vibrate and rotate.

Figure 1.3 shows the scattering mechanism when identical molecules are irradiated with monochromatic light, and also shows a diagram of the resulting spectrum. Two electronic levels—the ground state (0) and the first excited state (1)—and the excited vibrational levels are shown. The dashed lines denote two virtual levels of the molecule, separated by $h\nu_v$, where ν_v is the frequency of one of the possible normal vibrations of the molecule. According to the Boltzmann law, only a small proportion of the molecules will occupy the first excited vibrational state at room temperature. Most of them are in the ground state. If the molecules are now irradiated with monochromatic light, e.g. photons of energy $h\nu_0$, we observe a spectrum, part of which is shown schematically in Fig. 1.3b. In the centre we have the Rayleigh line of very strong intensity (log scale). The frequency ν_{Ray} that corresponds to it is equal to that of the incident light, ν_0. The intensity of the Rayleigh line is 10^3–10^4 times greater than that of the accompanying Raman bands of frequencies $\nu_{R(St)}$ (smaller than

ν_0) and $\nu_{R(aSt)}$ (greater than ν_0). The absolute differences between the frequencies of the incident photon and both scattered photons are the same and equal to the molecular vibration frequency (Fig. 1.3a):

$$h\nu_0 - h\nu_{R(St)} = h\nu_v \quad \text{and hence} \quad \nu_0 - \nu_{R(St)} = \nu_v$$

$$h\nu_{R(aSt)} - h\nu_0 = h\nu_v \quad \text{and hence} \quad \nu_{R(aSt)} - \nu_0 = \nu_v$$

We see now why the difference in frequency between the exciting radiation and the scattered radiation is characteristic of a molecule and independent of the frequency of the exciting radiation.

The intensity of the anti-Stokes Raman bands of frequency $\nu_{R(aSt)}$ is much lower than that of the Stokes bands in view of the difference of population of the ground and excited states in a set of molecules at room temperature, already mentioned. Therefore Raman spectrometry usually involves measurement of the Stokes band.

In order to explain the differences in the mechanisms underlying the three methods most often used today for investigating rotational–vibrational spectra, viz. infrared spectrometry, fluorimetry and Raman spectrometry, the ground- and excited-state potential energy (U) curves are presented in Fig. 1.4 together with the vibrational levels and the transitions between them observed by the three methods.

(i) For Raman scattering we have:

$$h\nu_v = h\nu_0 - h\nu_{R(St)}$$

where ν_0 is the incident beam frequency, and $\nu_{R(St)}$ is the measured frequency of the Stokes scattered radiation.

(ii) For infrared spectrometry:

$$h\nu_v = h\nu_0 = h\nu_{IR}$$

where ν_0 is the frequency of the beam incident on the sample and equal to ν_{IR}—the frequency of the radiation absorbed.

(iii) In fluorimetry:

$$h\nu_v = h\nu_0 - h\nu_{em}$$

where ν_0 is the frequency of the incident beam, and ν_{em} is the frequency of the radiation emitted.

1.3 POLARIZABILITY OF THE MOLECULE

The electric field of an electromagnetic wave shifts the centre of the negative charge in the molecule with respect to the centre of the positive charge (Fig. 1.5), *inducing a dipole moment* μ_{ind}, proportional to the electric field intensity **E**:

$$\boldsymbol{\mu}_{\text{ind}} = \boldsymbol{\alpha}\mathbf{E} \tag{1.1}$$

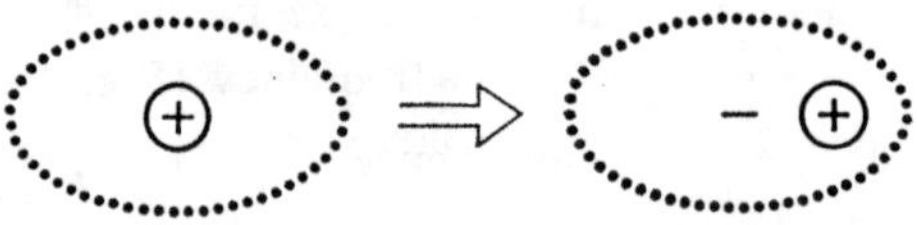

Fig. 1.5—Generation of the dipole moment induced by electric field of the electromagnetic wave.

The proportionality factor $\boldsymbol{\alpha}$, known as the *polarizability*, is characteristic of the molecule. The polarizability of a molecule is related to the mobility of its electrons and is always non-zero. It is a function of the shape and size of the molecule and has the dimensions of volume [cm^3]. The polarizability of a molecule is usually anisotropic, i.e. different for different directions in space. It is smaller for a single bond than for multiple bonds. There are, however, some isotropic molecules, for which the polarizability has the same value irrespective of the direction in space.

The variation of the polarizability with direction in space can be given most generally in the form of the polarizability tensor [12], i.e. a set of nine values α_{ij} (tensor components) related to the rectangular spatial coordinate system $(i, j = x, y, z)$:

$$\boldsymbol{\alpha}_{ij} = \begin{bmatrix} \alpha_{xx} & \alpha_{xy} & \alpha_{xz} \\ \alpha_{yx} & \alpha_{yy} & \alpha_{yz} \\ \alpha_{zx} & \alpha_{zy} & \alpha_{zz} \end{bmatrix}$$

The polarizability tensor is symmetrical, i.e. $\boldsymbol{\alpha}_{ij} = \boldsymbol{\alpha}_{ji}$. The properties of every tensor change according to the coordinate system selected. However, there are two values which are independent of the coordinates. These are known as *invariants* of the system and are referred to as the *mean value* and *anisotropy*. For the polarizability tensor these are: the mean polarizability $\bar{\alpha}$ determined as:

$$\bar{\alpha} = \tfrac{1}{3}(\alpha_{xx}+\alpha_{yy}+\alpha_{zz}),$$

where α_{xx}, α_{yy} and α_{zz} are the diagonal components of the polarizability tensor (see matrix $\boldsymbol{\alpha}_{ij}$), and the anisotropy γ given by the equation:

$$\gamma^2 = \tfrac{1}{2}[(\alpha_{xx}-\alpha_{yy})^2+(\alpha_{yy}-\alpha_{zz})^2+(\alpha_{zz}-\alpha_{xx})^2+6(\alpha_{xy}^2 + \alpha_{xz}^2+\alpha_{yz}^2)],$$

The diagonal values α_{xx}, α_{yy} and α_{zz} give the semi-axes of the polarizability ellipsoid. In the case of an isotropic molecule the polarizability ellipsoid becomes a sphere. The anisotropy γ is the measure of departure of the polarizability ellipsoid from spherical (regular) symmetry. In the case of spherical symmetry we have in the polarizability tensor only non-zero, equal diagonal components, the remaining components of the tensor being equal to zero, so the anisotropy is also equal to zero.

In Raman scattering the *tensor of the polarizability derivative with respect to the vibrational normal coordinate Q*

$$\alpha'_{ij} = \left(\frac{d\alpha}{dQ}\right)_0$$

is of great importance: the intensity of the Raman scattering depends on that tensor.

The polarizability derivative tensor can also be presented in the form of a matrix—a set of nine numbers, analogous to that for the polarizability tensor. The polarizability derivative tensor also has two invariants—values that are independent of the coordinate system. As for the polarizability tensor, they are the mean value $\bar{\alpha}'$ defined as:

$$\bar{\alpha}' = \tfrac{1}{3}(\alpha'_{xx}+\alpha'_{yy}+\alpha'_{zz})$$

and the anisotropy given by the formula:

$$(\gamma^*)^2 = \tfrac{1}{2}[(\alpha'_{xx}-\alpha'_{yy})^2+(\alpha'_{yy}-\alpha'_{zz})^2+(\alpha'_{zz}-\alpha'_{xx})^2 + 6(\alpha'^2_{xy}+\alpha'^2_{xz}+\alpha'^2_{yz})]$$

In the general case of an electric field with three components in the three directions x, y, z, vector **E** is described by three values: E_x, E_y and E_z. The expanded formula (1.1) has the form:

$$\mu_{\text{ind},x} = \alpha_{xx}E_x+\alpha_{xy}E_y+\alpha_{xz}E_z$$
$$\mu_{\text{ind},y} = \alpha_{yx}E_x+\alpha_{yy}E_y+\alpha_{yz}E_z$$
$$\mu_{\text{ind},z} = \alpha_{zx}E_x+\alpha_{zy}E_y+\alpha_{zz}E_z$$

1.4 THE MOLECULE AS A VIBRATING DIPOLE, AND THE CLASSICAL DESCRIPTION OF RAMAN SCATTERING

The classical interpretation of light scattering assumes that the molecule is a vibrating dipole. That dipole is generated when an electromagnetic wave with a periodically varying intensity of the electric component (E) interacts with the molecule:

$$\mathbf{E} = \mathbf{E}_0\cos(2\pi\nu_0 t) \tag{1.2}$$

where $\mathbf{E}_0$ is the intensity amplitude of the electric component, ν_0 is the wave frequency, and t is time.

According to the principles of electrodynamics, every oscillating dipole becomes a source of an electromagnetic wave of frequency equal to the frequency of oscillation of the dipole. That wave is propagated in space in all directions except along the dipole axis (i.e. the direction of the induced dipole moment).

Substituting (1.2) into (1.1) we obtain the equation determining the value of the dipole moment induced by the electromagnetic wave:

$$\boldsymbol{\mu}_{\mathrm{ind}} = \boldsymbol{\alpha}\mathbf{E}_0 \cos(2\pi\nu_0 t) \tag{1.3}$$

That value varies in accordance with the wave frequency and the molecule becomes the source of radiation of that frequency (ν_0). In this way we obtain the Rayleigh scattering, which would be the only possible form of scattering if the molecule did not have its own internal vibrations. In the course of vibration, the molecule may change its size and shape, and thus its polarizability (the shape of the polarizability ellipsoid). For a molecule vibrating with a certain frequency of its own, every component of the polarizability tensor α_{ij} may be presented in the approximate form:

$$\alpha_{ij} = (\alpha_{ij})_0 + \left(\frac{\partial \alpha_{ij}}{\partial Q}\right)_0 Q \tag{1.4}$$

where $(\partial\alpha_{ij}/\partial Q)_0$ is the change of component α_{ij} in the course of the given vibration, characterized by the normal coordinate Q, describing the displacement of all nuclei of the atoms in the molecule during vibration about their equilibrium positions, and $(\alpha_{ij})_0$ is the value of α_{ij} at the equilibrium position of the nuclei. Lack of knowledge of the function relating polarizability to the normal coordinates of the molecule vibrations obliges us to present the value α_{ij} in an approximate form. Thus the approximate values of the function, i.e. of α_{ij}, are found by expanding the function into a Maclaurin series. In doing this we assume that the displacements of the nuclei are close to zero. A further approximation consists in preserving only the first two terms of that series (harmonic approximation).

The normal coordinate varies periodically during vibration, as indicated by the equation:

$$Q_\nu = \mathbf{A}_\nu \cos(2\pi\nu t) \tag{1.5}$$

where $\mathbf{A}_\nu$ is the amplitude of the given normal vibration of the molecule of frequency ν, and t is time.

Presenting the polarizability tensor in the form of a matrix we write:

$$(\alpha_{ij}) = \boldsymbol{\alpha} \quad \text{(where } i, j = x, y, z)$$

Considering (1.4) and (1.5), we can write for the case when the molecule has normal vibrations of frequency ν:

$$\boldsymbol{\alpha} = \boldsymbol{\alpha}_0 + \left(\frac{\partial \boldsymbol{\alpha}}{\partial Q}\right)_0 \mathbf{A}_\nu \cos(2\pi\nu t) \tag{1.6}$$

Substituting this expression for $\boldsymbol{\alpha}$ into Eq. (1.3) we obtain an expression describing the value of the moment induced in a molecule with a vibration of frequency ν, acted upon by an electromagnetic wave of frequency ν_0:

$$\boldsymbol{\mu}_{\mathrm{ind}} = \boldsymbol{\alpha}_0 \mathbf{E}_0 \cos(2\pi\nu_0 t) + \left(\frac{\partial \boldsymbol{\alpha}}{\partial Q_\nu}\right)_0 \mathbf{A}_\nu \mathbf{E}_0 \cos(2\pi\nu_0 t)\cos(2\pi\nu t) \tag{1.7}$$

Bearing in mind that $\cos\alpha\cos\beta = \frac{1}{2}\cos(\alpha-\beta)+\frac{1}{2}\cos(\alpha+\beta)$ we finally get:

$$\boldsymbol{\mu}_{\mathrm{ind}} = \boldsymbol{\alpha}_0 \mathbf{E}_0 \cos(2\pi\nu_0 t) + \frac{1}{2}\left(\frac{\partial \boldsymbol{\alpha}}{\partial Q_\nu}\right)_0 \mathbf{A}_\nu \mathbf{E}_0 \{\cos[2\pi(\nu_0-\nu)t] + \cos[2\pi(\nu_0+\nu)t]\} \tag{1.8}$$

The vibrating molecule can therefore be the source of scattered radiation of three different frequencies: (i) of frequency ν_0, i.e. frequency unchanged with respect to the incident radiation (Rayleigh scattering), (ii) of frequency $(\nu_0-\nu)$, i.e. frequency equal to the difference between the incident radiation frequency and that of the vibrations of the molecule (the Raman Stokes scattering) and (iii) of frequency $(\nu_0+\nu)$, i.e. of a frequency equal to the sum of the frequencies of the incident radiation and the vibration of the molecule (Raman anti-Stokes scattering).

1.5 RAMAN SCATTERING INTENSITY

In the early thirties of the present century Placzek [13] developed a polarizability theory that allowed the formulation of equations determining the experimentally observed intensities of the Raman Stokes and anti-Stokes scattering. The assumptions of that theory required that the frequency ν_0 of the exciting radiation be much higher than that of vibration ν_v of the molecule $(\nu_0 \gg \nu_v)$ and that the frequency ν_{el} corresponding to the electron transition of the molecule from the ground state to the first excited state be much higher than that of the exciting radiation $\nu_0(\nu_{el} \gg \nu_0)$. Because of this latter requirement the *Placzek theory* finds application only in the case of normal Raman scattering and does not encompass the resonance Raman scattering (see Section 1.8). The Placzek theory provides Eqs.

(1.9) and (1.10) [14] describing the intensities of the Stokes ($I^{St}_{\nu,ij}$) and anti-Stokes ($I^{aSt}_{\nu,ij}$) radiations, calculated per unit intensity of the exciting radiation and scattered by one molecule or unit cell of a crystal vibrating with its own frequency ν. The equations have been derived for scattered radiation observed at an angle of 90° to the exciting beam, assuming that the directions of the electric vectors of the exciting and observed scattered radiations are in accordance with the respective axes of the rectangular coordinate system assigned to the molecule or unit cell of a crystal.

$$I^{St}_{\nu,ij} = \frac{2\pi^2 h}{c} \frac{(\tilde{\nu}_0 - \tilde{\nu})^4}{\tilde{\nu}[1-\exp(-hc\tilde{\nu}/kT)]} g_\nu \left(\frac{\partial \alpha_{ij}}{\partial Q_\nu}\right)_0^2 \tag{1.9}$$

$$I^{aSt}_{\nu,ij} = \frac{2\pi^2 h}{c} \frac{(\tilde{\nu}_0 + \tilde{\nu})^4}{\tilde{\nu}[\exp(hc\tilde{\nu}/kT)-1]} g_\nu \left(\frac{\partial \alpha_{ij}}{\partial Q_\nu}\right)_0^2 \tag{1.10}$$

where $\tilde{\nu}_0$ and $\tilde{\nu}$ are wave numbers (cm^{-1}) corresponding to the frequencies of the exciting radiation and the molecular vibration, respectively, T is the absolute temperature of the sample, h is the Planck constant, c is the velocity of light, g_ν is the degree of degeneracy of the given vibration of the molecule, $(\partial\alpha_{ij}/\partial Q_\nu)_0$ is the change of the α_{ij} component of the polarizability tensor for that vibration with respect to the normal coordinate Q_ν. The intensity of the scattered radiation per molecule is called the *scattering coefficient* and is characteristic of the given vibration of the molecule under constant excitation and measurement conditions.

As seen from the equations above, the *intensity of the scattered Raman radiation is proportional to the square of the polarizability derivative with respect to the normal vibration coordinate,* i.e. it depends on the variation of polarizability during vibration. It is also proportional to the fourth power of the scattered-vibration frequency.

From Eqs. (1.9) and (1.10) we can calculate the anti-Stokes to Stokes radiation intensity ratio:

$$\frac{I^{aSt}}{I^{St}} = \frac{(\tilde{\nu}_0 + \tilde{\nu})^4}{(\tilde{\nu}_0 - \tilde{\nu})^4} e^{-h\tilde{\nu}/kT}$$

As can be seen, the ratio depends on the absolute temperature of the sample.

1.6 DEPOLARIZATION RATIOS

The direction of the dipole moment induced by the electric vector of the electromagnetic wave interacting with a molecule depends on the symmetry of the molecule. From that symmetry results the isotropic or anisotropic spatial distribution of the molecular polarizability.

During a vibration that is active in Raman scattering, a change of polarizability of the molecule occurs. The derivative of polarizability with respect to the normal coordinate is non-zero: $(\partial \boldsymbol{\alpha}/\partial Q)_0 \neq 0$. The spatial distribution of the polarizability derivative characteristic of the given vibration is isotropic or anisotropic depending on whether the *vibration* is *fully symmetric* (with respect to all elements of symmetry of the molecule) or *antisymmetric* (with respect to any element of symmetry of the molecule), and on the type of symmetry of the molecule itself. In the case of a fully symmetric vibration of a molecule of highest symmetry (spherical) the spatial distribution of the polarizability derivative is isotropic (a). In all other cases the distribution is anisotropic (b). In case (a) the direction of the electric vector of the Raman-scattered radiation is in accordance with the direction of the electric vector of the exciting beam, in case (b) it is in disaccordance with that direction.

We shall now discuss both cases, considering two methods of sample illumination: (i) with a linear polarized exciting beam (e.g. laser beam), (ii) with an unpolarized exciting beam (e.g. conventional mercury lamp illumination). The scattered Raman radiation is observed at an angle of 90° with respect to the exciting beam. Two polaroid plates are used successively, to transmit polarized radiation: one in the vertical direction of the electric vector (z-axis direction), in which intensity $I_{\parallel}$ is observed, and the second in the horizontal direction of the electric vector (y-axis direction), in which intensity $I_{\perp}$ is observed.

If the vibration of a molecule of spherical symmetry is fully symmetric (Fig. 1.6) and type (i) illumination is applied, we observe successively intensity $I_{\parallel} > 0$ and $I_{\perp} = 0$. In the scattered radiation only the vertical component $\mathbf{E}_V$ occurs (i.e. the scattered radiation is linearly polarized). The direction of the electric vector is identical in the exciting and scattered radiations. The *depolarization ratio* ϱ is defined as

$$\varrho = \frac{I_{\perp}}{I_{\parallel}} \tag{1.12}$$

and equals zero in the case considered.

If illumination of type (ii) is used, we observe the intensities $I_{\parallel} > 0$ and $I_{\perp} = 0$, so ϱ equals 0. The scattered radiation is linearly polarized and has the direction of the electric vector $\mathbf{E}_V$, though the exciting radiation is not polarized. This comes from the fact that all the directions of the electric vectors of the non-polarized exciting beam lie in a plane parallel to the xz plane. Thus the observed direction of the electric vector in the scattered beam—being a projection of any vector in the plane xz onto the z-axis—is in accordance with the direction $\mathbf{E}_V$, irrespective of the direction of the electric vector of the exciting beam.

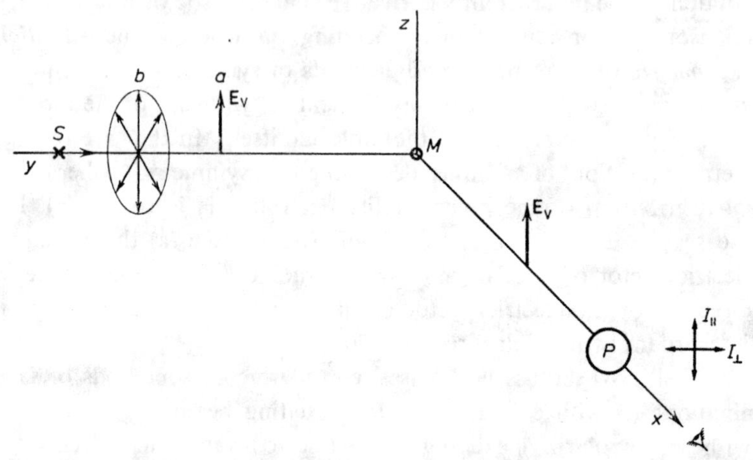

Fig. 1.6—Depolarization ratio of the band corresponding to totally symmetric vibration of a molecule of spherical symmetry. $I_{\perp} = 0$, $I_{\parallel} > 0$, $\varrho = I_{\perp}/I_{\parallel} = 0$. (*a*) Linearly polarized light. (*b*) Unpolarized light. (*S*) Source of light. (*M*) Scattering molecule. (*P*) Polarization analyser.

In the case of totally symmetric vibrations of molecules with symmetry other than spherical, and of any antisymmetric vibrations, we observe anisotropy of the spatial distribution of the polarizability derivative. The direction of the electric vector **E** of the scattered radiation (Fig. 1.7) is in disaccordance with the direction of the electric vector of the exciting beam, and depends on the position of the molecule with respect to the direction of the electric vector of the excitation beam. In the scattered radiation observed at 90° to the direction of the excitation beam, we detect two components of the radiation (by using the polaroid plates described above), with the electric vector directions $\mathbf{E}_V$ and $\mathbf{E}_H$. Here neither of these components is equal to zero.

It follows from theoretical considerations that if the Raman Stokes radiation scattered by a gaseous or liquid sample is observed at right-angles to the unpolarized exciting beam, then when a scattered radiation analyser is used (Figs. 1.6b and 1.7b), Eq. (1.9) assumes one of the following forms:

$$I_{\perp} = \frac{2\pi^2 h}{c} \frac{(\tilde{\nu}_0 - \tilde{\nu})^4}{\tilde{\nu}[1 - \exp(-hc\tilde{\nu}/kT)]} g \frac{6}{45} \gamma^{*2} \tag{1.13}$$

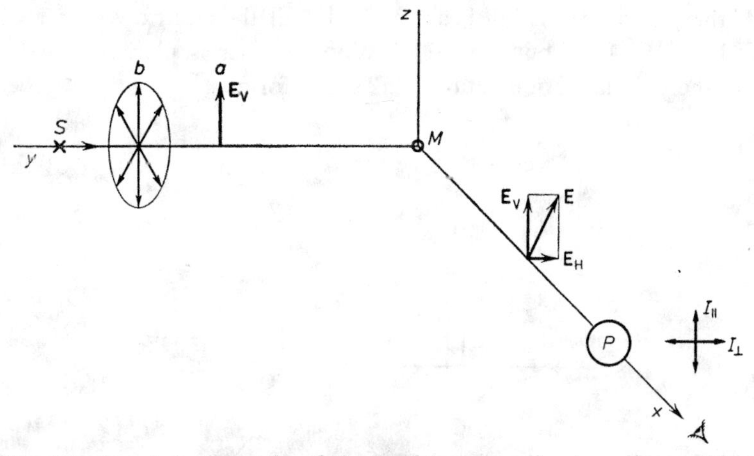

Fig. 1.7—Depolarization ratio of the band corresponding to antisymmetric or degenerate vibration or totally symmetric vibration of molecules of non-spherical symmetry. $I_\perp > 0$, $I_{||} > 0$, $\varrho = I_\perp/I_{||} > 0$. (E) Direction of the electric vector of the radiation scattered in direction x. Remaining notation as in Fig. 1.6.

$$I_{||} = \frac{2\pi^2 h}{c} \frac{(\tilde{\nu}_0-\tilde{\nu})^4}{\tilde{\nu}[1-\exp(-hc\tilde{\nu}/kT)]} g\left(\overline{\alpha}'^2+\frac{7}{45}\gamma^{*2}\right) \tag{1.14}$$

where $I_\perp$ is the intensity (per molecule) of the Raman scattering component with the electric vector $\mathbf{E}_H$ parallel to the excitation beam direction (y) (Fig. 1.7b), $I_{||}$ is the intensity of the component with the electric vector $\mathbf{E}_V$ perpendicular to the exciting beam direction; the remaining symbols are the same as in Eqs. (1.9) and (1.10), the subscript ν being omitted from all symbols characteristic of the considered vibration of the molecule (i.e. g, $\overline{\alpha}'$ and γ^*). In the equations above, two invariants of the polarizability derivative tensor occur: $\overline{\alpha}'$ and γ^*, which have been discussed earlier (Section 1.3).

The depolarization ratio $\varrho_n = I_\perp/I_{||}$ calculated from Eqs. (1.13) and (1.14) is

$$\varrho_n = \frac{6\gamma^{*2}}{45\overline{\alpha}'^2+7\gamma^{*2}} \tag{1.15}$$

An identical equation describing the depolarization ratio is obtained if the intensity of the scattered radiation is measured without using the polarization analyser but the exciting radiation is linearly polarized (Fig. 1.8): first in the direction of the electric vector $\mathbf{E}_V$ ($\mathbf{E}_V \perp x$) and secondly in the direction of $\mathbf{E}_H$ ($\mathbf{E}_H || x$).

If the polarized exciting beam is used with the electric vector $\mathbf{E}_V \perp x$, (Figs. 1.6a and 1.7a) but the polarization analyser is placed behind the sample, the depolarization ratio ϱ_s takes the form:

$$\varrho_s = \frac{3\gamma^{*2}}{45\bar{\alpha}'^2 + 4\gamma^{*2}} \tag{1.16}$$

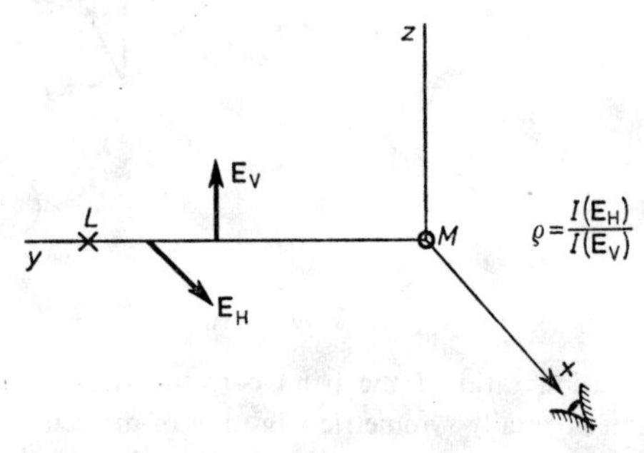

Fig. 1.8—Depolarization ratio measurement. (L) Laser. (M) Sample. ($\mathbf{E}_V$, $\mathbf{E}_H$) Directions of the electric vector of the exciting beam. ($I(\mathbf{E}_V)$) Measured intensity of Raman Stokes scattering when the sample is illuminated with the linearly polarized exciting beam with electric vector direction $\mathbf{E}_V$.($I(\mathbf{E}_H)$) Intensity of the scattered radiation in the case of an exciting beam with electric vector direction $\mathbf{E}_H$.

This is the most common method of measuring the depolarization ratio by use of Raman laser spectrometers. Equations (1.15) and (1.16) are true only when the scattered radiation is observed precisely at right-angles to the exciting beam direction. In the case of laser excitation sources, departures from this requirement (divergence of the exciting and scattered beams) are relatively small and thus the deviations of the measured values from the true ones are not so significant as in the case of conventional sources of excitation.

The values of $\bar{\alpha}'$ and γ^* depend on the symmetry of the molecule itself as well as on the given vibration. In the case of a molecule with spherical symmetry and a totally symmetric vibration, $\gamma^* = 0$, so both ϱ_n and ϱ_s are equal to zero [Eqs. (1.15) and (1.16)]. In the case of an antisymmetric vibration (with respect to any symmetry element of the molecule) or a degenerate vibration, $\bar{\alpha}' = 0$, and than $\varrho_n = 6/7$ and $\varrho_s = 3/4$. Thus totally symmetric vibrations are characterized by values of the degree of depolarization lying within the limits:

$$0 \leqslant \varrho_n < 6/7 \quad \text{or} \quad 0 \leqslant \varrho_s < 3/4$$

depending on the method used for measuring the depolarization ratio.

1.7 SELECTION RULES FOR VIBRATIONAL TRANSITIONS ACTIVE IN INFRARED AND RAMAN SPECTROMETRY

Infrared absorption spectrometry together with Raman spectrometry will give almost complete information about the vibrational spectrum of a molecule in the ground electronic state. The methods are complementary, as follows from the very nature of the phenomena on which they are based. They are also complementary because of the differing nature of the measuring techniques involved and their specific usefulness in different structural and analytical problems.

Both the infrared and the Raman spectra constitute sets of vibrational–rotational bands corresponding to the combination of vibrations of atoms in the molecule and rotations of the molecule itself. The number, frequency and amplitude of the vibrations characterize the molecule precisely. The two spectra, consisting of bands of characteristic contours and position, reflect faithfully the vibrational possibilities of the molecule.

The complementarity of infrared and Raman spectrometry results from the different *selection rules* which determine the appearance in the infrared and/or Raman spectrum of a band corresponding to a given vibration of the molecule (Table 8.2). *If the vibration causes a change in the dipole moment*, which happens when the vibration changes the symmetry of charge density distribution, i.e. if $(\partial\mu/\partial Q)_0 \neq 0$, *it is active in the infrared spectrum. If the vibration produces a change of the molecular polarizability*, i.e. if $(\partial\alpha/\partial Q)_0 \neq 0$, *it is active in the Raman spectrum.*

The fulfilment of one or both of these conditions is related to the symmetry of the molecule. There even exists the *rule of mutual exclusion*, stating that if there is a centre of symmetry in the molecule, a vibration that is active in the infrared spectrum is inactive in the Raman spectrum and vice versa. In the case of molecules without a centre of symmetry a number of the vibrations appear in both spectra. They often differ, however, as regards their relative intensity (Table 8.3), for instance the vibrations of strongly polar functional groups are more readily observed in the infrared spectrum, while the vibrations of double and triple bonds and the carbon-skeleton of the molecule are better seen in the Raman spectrum. Totally symmetric vibrations are only (or much better) visible in the Raman spectrum.

Why are there such differences in the selection rules and the recorded intensities of the infrared and Raman bands corresponding to the same vibration? The explanation comes from consideration of the behaviour of an anharmonic oscillator, used as a model of the vibrations of a molecule.

A molecular oscillator interacts with radiation and passes from energy state n to state m if certain conditions are fulfilled. The condition significant for our present considerations can be formulated as follows: the transition of a molecule from one energy state to another can take place when the probability of that transition is not equal to zero. The transition probability in the case of infrared absorption is proportional to the square of the absolute value of the so-called *transition moment*, which is given by the equation:

$$|M_{nm}| = \int_{-\infty}^{+\infty} \Psi_n^* \mu \Psi_m \mathrm{d}Q \tag{1.17}$$

where Ψ_n and Ψ_m are the wave functions of states n and m, the dipole moment of the molecule being the transition operator, and the integration carried out over the normal coordinate Q of the vibration. Therefore if there is no change of the dipole moment over the normal coordinate of the vibration, the transition moment equals zero and the transition is inactive in the infrared spectrum.

The measured integral absorbance of the infrared band is proportional to the square of the dipole moment derivative along the normal coordinate of the vibration. It therefore depends also on the magnitude of variation of the dipole moment during vibration.

In the case of the Raman spectrum the transition moment is given by the expression [15]:

$$|(\alpha_{ij})_{nm}| = \int_{-\infty}^{+\infty} \Psi_n^* \alpha_{ij} \Psi_m \mathrm{d}\tau \tag{1.18}$$

Here polarizability is the transition operator and integration is made over the volume element $\mathrm{d}\tau$, since polarizability is a tensor and not a vector like the dipole moment. The measured integral intensity of the Raman band is proportional to the square of the polarizability derivative along the normal coordinate and therefore depends on the magnitude of polarizability variations during vibration.

The question therefore arises, why during the same vibration only the polarizability or only the dipole moment will change, since the values of both these quantities depend on the possibility of displacement of the nuclei and electrons in the molecule. It would seem that this necessarily leads to the simultaneous activity of vibration in both spectra. We must remember, however, that the variation of the dipole moment or polarizability responsible for the vibration activity in the given spectrum relates to the molecule as a whole and not to its particular elements. For this reason the symmetry of the molecule and the symmetry of vibration are decisive for the activity of a vibration (see Chapter 2).

The CO_2 molecule is a simple example of accordance of the given selection rules with experiment. That molecule has four degrees of freedom and thus four normal vibrations: ν_1—symmetrical stretching vibration, ν_3—antisymmetrical stretching vibration, $\nu_{2,4}$—bending vibration, doubly degenerate. In Fig. 1.9 the variation of the polarizability and dipole moment

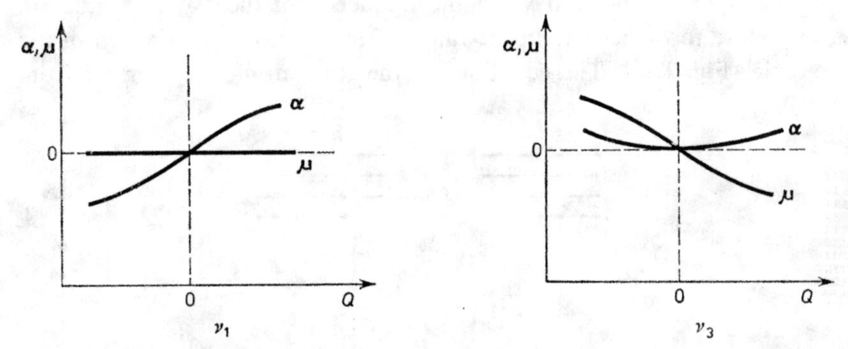

Fig. 1.9—Change of polarizability and dipole moment of the CO_2 molecule during symmetric—ν_1 and antisymmetric—ν_3 stretching vibrations.

during vibrations ν_1 and ν_3 are presented for the CO_2 molecule as a whole. The variations of the dipole moment and polarizability for the molecule as a whole (and not for the individual bonds), in the cases of the particular vibrations, are as follows:

$$(\nu_1) \quad \overleftarrow{\mathrm{O}}{=}\mathrm{C}{=}\overrightarrow{\mathrm{O}} \quad \left(\frac{\partial \mu}{\partial Q}\right)_0 = 0 \quad \left(\frac{\partial \alpha}{\partial Q}\right)_0 \neq 0$$

$$(\nu_3) \quad \overleftarrow{\mathrm{O}}{=}\overrightarrow{\mathrm{C}}{=}\overrightarrow{\mathrm{O}} \quad \left(\frac{\partial \mu}{\partial Q}\right)_0 \neq 0 \quad \left(\frac{\partial \alpha}{\partial Q}\right)_0 = 0$$

$$(\nu_{2,4}) \quad \begin{array}{c} \mathrm{O}{=}\overset{\uparrow}{\mathrm{C}}{=}\mathrm{O} \\ \downarrow \qquad \downarrow \\ \mathrm{O}{=}\mathrm{C}{=}\mathrm{O} \\ \oplus \quad \ominus \quad \oplus \end{array} \quad \left(\frac{\partial \mu}{\partial Q}\right)_0 = 0 \quad \left(\frac{\partial \alpha}{\partial Q}\right)_0 \neq 0$$

The vibrations ν_3 and $\nu_{2,4}$ are active only in the infrared spectrum, while the ν_1 vibration is active only in the Raman spectrum. The CO_2 molecule has a symmetry centre and, as is seen, the rule of mutual exclusion is fulfilled.

1.8 THE RESONANCE RAMAN EFFECT

The application of laser excitation sources has not only contributed to the revival of Raman spectrometry but also to the discovery of a number of new effects which could only be observed thanks to the high power of these sources. These effects are now the object of research in physics.

In the present book we shall limit our considerations to the resonance Raman effect (RRE) since it has already found application in solving problems in biochemistry and chemical analysis.

The RRE is observed when the frequency of the exciting radiation is very close to or lies within the range of an electronic absorption of the molecule (Fig. 1.10). The equations of the Raman intensity derived from

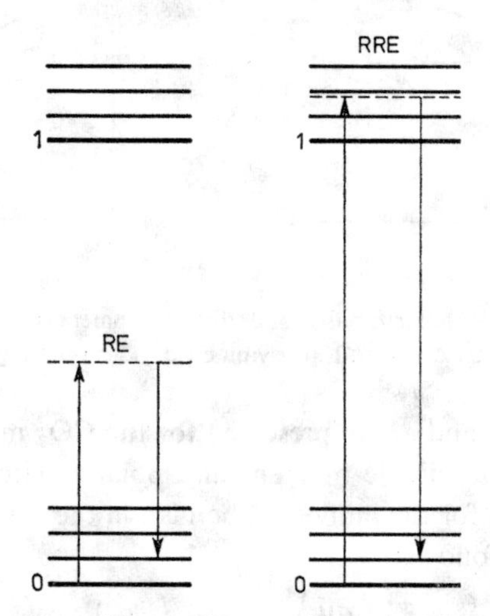

Fig. 1.10—Diagram of transitions of Raman (Stokes) scattering (RE) and resonance Raman (Stokes) scattering (RRE). (0) Ground electronic state, (1) First excited electronic state.

the Placzek theory do not encompass the latter case. The RRE was first observed at the time when only conventional sources of light were available [17], but it was only the strongly monochromatic laser beam which made simple observation of the effect possible.

The Raman intensity is, according to known theories [16–18], proportional to an expression with the difference $(\nu_e - \nu_0)$ in the denominator (where ν_e is the frequency of the electronic transition of the molecule from the ground to the first excited state, and ν_0 is that of the exciting radiation). Hence the intensity increases significantly when that difference tends to zero. The very high intensity of RRE makes it particularly useful for chemical analysis. The detectability of a component in a sample, so far achieved in infrared absorption and conventional Raman scattering, is very rarely as low as a few tenths of 1%. The use of RRE makes it possible to improve the detectability by several orders of magnitude.

The theory of the resonance Raman effect has not as yet been developed to the point where it could give a clear answer to the question: what is the essential difference between that effect and resonance fluorescence? Behringer and Hester [11] give two features that distinguish resonance fluorescence from RRE, viz. the quenching of the fluorescence with time (not observed in the case of scattering) and the difference in dependence of the measured intensity on the number of molecules interacting with the exciting beam. The Raman intensity is directly proportional to the number of scattering molecules, i.e. to the volume concentration of the substance under constant measuring conditions (see Chapter 5). The latter fact is significant for analytical applications of Raman spectrometry.

Use of the RRE in structural analysis requires the application of laser excitation radiation of a frequency appropriate for the given molecule. That frequency must lie within the range of electronic absorption by the molecule. Most suitable sources for this purpose are tunable dye lasers. Colourless substances require lasers generating frequencies in the ultraviolet region. In modern Raman spectrometers, He–Ne and Ar^+ gas lasers are most often used. They allow the generation of only a few lines, practically all in the visible region, and thus can only be used for recording the RRE of coloured samples.

In RRE studies there is always the chance that strong absorption of the exciting radiation may occur and cause decomposition of the sample. Also the Raman-scattered radiation may be absorbed by the sample itself. If fluorescence appears simultaneously with RRE, it produces a strong background in the recorded RR spectra.

The above-mentioned features of RRE cause some difficulties in the analytical application of this effect. These difficulties are reduced by using highly diluted samples (concentrations of the order of 10^{-5}–10^{-7} mole/l.) and rotating cells or targets (cf. Fig. 3.14, p. 70). It should also be noted that the laser beam should be positioned in the sample cell so that there is only a thin layer of sample between it and the monochromator slit. It is easiest to select the appropriate frequency of the excitation beam by recording the absorption and fluorescence spectra of the test sample (Fig. 1.11). Best results, i.e. the clearest RR spectra (Fig. 1.12), are obtained if

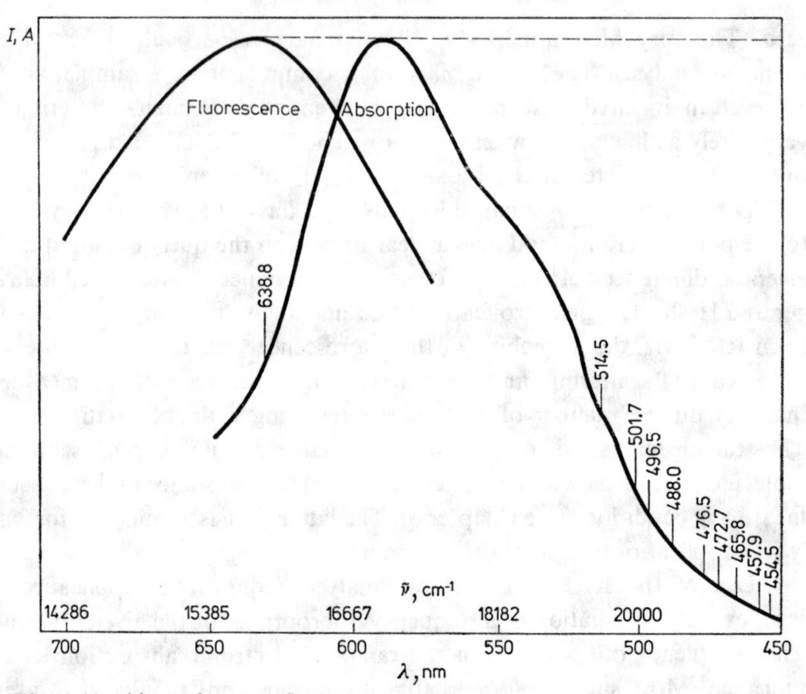

Fig. 1.11—Absorption and fluorescence spectra of $5\times10^{-6}M$ aqueous Crystal Violet solution, with the He–Ne and Ar^+ laser lines marked. The values of fluorescence intensity (I) and absorbance (A) have been normalized to the maxima as unity (test at the Institute of Industrial Chemistry, Warsaw, 1975, by H. Barańska, A. Łabudzińska and W. Andrzejewska).

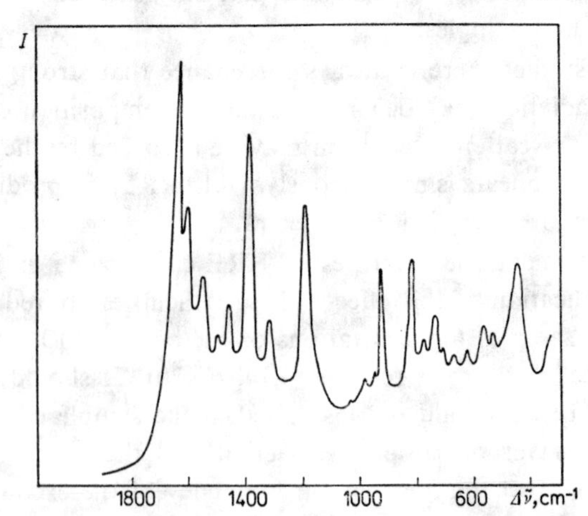

Fig. 1.12—Resonance Raman spectrum of $5\times10^{-6}M$ aqueous Crystal Violet solution with excitation at 514.5 nm (Ar^+ laser) (test as in Fig. 1.11).

the frequency of the excitation beam lies as near as possible to the maximum of the absorption band and as far as possible from the maximum of the fluorescence band of the sample. The contour of the electronic absorption band is significantly affected by any solvent (liquid) or diluent (solid) added to the test compound.

RRE has found wide application in the study of aqueous solutions of composite biological compounds (see Chapter 7).

REFERENCES

[1] A. Smekal, *Naturwiss.*, 1923, **11**, 873.
[2] H. A. Kramers and W. Heisenberg, *Z. Phys.*, 1925, **31**, 681.
[3] E. Schrödinger, *Ann. Phys. (Leipzig)*, 1926, **81**, 109.
[4] P. A. M. Dirac, *Proc. Roy. Soc.*, 1927, **114**, 710.
[5] C. V. Raman, *Nature*, 1928, **121**, 501, 619, 721.
[6] G. Landsberg and L. Mandelstam, *Naturwiss.*, 1928, **16**, 557, 772.
[7] Y. Rocard, *Compt. Rend.*, 1928, **186**, 1107.
[8] J. Cabannes, *Compt. Rend.*, 1928, **186**, 1201.
[9] S. P. S. Porto and D. L. Wood, *Opt. Soc. Am.*, 1962, **52**, 251.
[10] J. Behringer, *The Relation of Resonance Raman Scattering to Resonance Fluorescence*. Manuscript made available courtesy of the author.
[11] *Molecular Spectroscopy*, Vol. 2, The Chemical Society, London, 1974.
[12] R. P. Feynman, R. B. Leighton and M. Sands, *The Feynman Lectures on Physics*, Addison-Wesley, Reading, Massachusetts, 1963–1965.
[13] G. Placzek, *Rayleigh-Streuung und Raman-Effekt*, in *Handbuch der Radiologie*, E. Marx (ed.), Part II, Vol. 6, Akademische Verlag, Leipzig, 1934.
[14] B. Schrader, *Angew. Chem. Int. Ed. Engl.*, 1973, **12**, 884.
[15] Z. Kęcki, *Fundamentals of Molecular Spectroscopy*, 2nd Ed., PWN, Warsaw, 1975.
[16] P. P. Shorygin, *Izv. Akad. Nauk. SSR, Ser. Fiz.*, 1948, **12**, 576; *Usp. Fiz. Nauk*, 1973, **109**, 293.
[17] J. Brandmüller and H. Moser, *Einführung in die Ramanspektroskopie*, Steinkopff Verlag, Darmstadt, 1962.
[18] H. J. Bernstein, *Resonance Raman Spectra*, 1974. Manuscript made available courtesy of the author.

2

Elements of group theory

A. Łabudzińska

2.1 MOLECULAR VIBRATIONS

Every molecule can vibrate only with certain characteristic frequencies determined by its spatial structure. To determine accurately the relationship between the bands in the spectrum and the corresponding molecular vibrations it is necessary to assume a spatial model of the molecule and identify all the vibrations that the molecule, as described by the model, can perform. Considerations of this type are the subject of structural analysis. From a study of the different models for molecules and the systems of bands in the spectra predicted for them, we can select a model, for any particular molecule, which will fit the recorded spectrum.

The molecular structure is characterized by a *set of symmetry elements* contained in the spatial model. The vibrations that the molecule can perform are determined by *symmetry operations* related to these elements. In mathematical terms it is assumed that the set of symmetry operations belonging to the given molecule forms a *group*. In the case of vibrational motions of the molecule (i.e. those taking place without displacement of the whole molecule in space) we are concerned with a *point group*, which means that during these vibrations at least one point in the molecule (the centre of symmetry) remains immobile.

The symmetry of the molecule and the vibrations it imposes are described by the *theory of point groups*. Calculation of the number of vibrations of the given molecule and distinguishing between those producing changes in dipole moment and those producing changes in polarizability of the molecule lead to the so-called *selection rules*, i.e. to the determination of the number of vibrational bands in the infrared absorption spectra and the Raman scattering spectra.

In the present chapter fundamentals of the point group theory are given, but the laws and equations are not derived. The method of applying the theory in molecular spectroscopy [1–3] is also presented.

2.2 ELEMENTS OF SYMMETRY

The molecular symmetry is determined by symmetry elements that characterize the molecule in the equilibrium state configuration. It has been found that five initial symmetry elements are sufficient to construct a symmetry model of any molecule. Some of those elements are included in the symmetry model of any given molecule.

The symmetry elements can be defined in the simplest way by the corresponding *symmetry operations*, i.e. displacements of an object that bring it into a coincidence position indistinguishable from the initial position. The symmetry element is the geometric object (e.g. point, line, plane) with respect to which the symmetry operation is performed.

The symmetry elements used to characterize molecules are:
identity element (E),
symmetry axis (C_n),
symmetry plane (σ),
centre of symmetry (i),
rotation–inversion axis (S_n).

Identity element (E). This element means that the molecule is identical to itself. It corresponds to the symmetry operation by means of which every atom returns to its initial position (i.e. it remains at rest). Without introduction of the identity element a mathematical approach to the symmetry of molecules in terms of group theory would not be possible. This element is present in all molecules.

Symmetry axis (C_n). The rotation of a molecule about the symmetry axis by a given angle yields a configuration that does not differ in any respect from the initial configuration; this means that identical atoms have changed places. The rotation angle fulfilling that condition is denoted as $360°/n$. The index n, called the *multiplicity of the axis*, takes the values $n = 1, 2, 3, ...$ (in practice $n = 1, 2, ..., 8$ and ∞). The single-fold axis C_1 means that only a full rotation by 360° brings the molecule back to its initial form; an infinite-fold axis C_∞ means that the molecule preserves its configuration after rotation by any angle, which may be infinitely small. A rule that is assumed is that in a model with several axes of symmetry the one of greatest multiplicity is vertical.

Symmetry plane (σ). The appearance of a symmetry plane in the model of a molecule means that the spatial arrangement of atoms lying on one side of that plane is a mirror image of the arrangement on the other side. The symmetry plane containing the symmetry axis of greatest multiplicity, and the plane perpendicular to that axis are denoted by σ_v and σ_h, respectively. The symbol σ_d denotes the diagonal plane, i.e. one including the rotation axis and bisecting the angle between two other axes.

Centre of symmetry (i), also known as the *centre of inversion*. The molecule has a centre of symmetry if changing the coordinates (x, y, z) of every atom to $(-x, -y, -z)$ yields a configuration equivalent to the initial one. The centre of symmetry is the origin of the coordinates. There are very few molecules which have a centre of symmetry as the only element of symmetry (apart from identity).

Rotation–inversion axis (S_n). The symmetry operation connected with this symmetry element is the sum of two operations: rotation about the axis by an angle of $360°/n$ and reflection from a plane perpendicular to that axis.

Every molecule has some elements of symmetry. Usually there are several, but there must be at least one, the identity element. A fully asymmetric molecule is not susceptible to any symmetry operations other than that of identity. Depending on their structure, molecules have various combinations of symmetry elements. The numbers of such different combinations, constituting sets of symmetry elements encountered in molecules, is limited and, as it appears, not very large.

Group theory gives a mathematical description of the behaviour of objects susceptible to definite symmetry operations. In the case of molecules we deal with the point group theory, since in all the symmetry operations to which molecules are amenable, at least one point in the spatial model of the molecule is at rest. For the symmetry of crystals, the space group theory applies, as here the possibility of a whole unit cell being displaced to another position in the crystal should be taken into account.

2.3 POINT GROUPS

It follows from group theory that a set of elements forms a point group if it fulfils the following four conditions.

1. The product of two elements of the set (A and B) is also an element of the set (C), thus $AB = C$.

2. The product of elements is associative: $A(BC) = (AB)C$.

3. The set of elements which forms the group contains element E, such that for every element C we have $EC = CE = C$. This element E is the identity element. We say that element E commutes with every element C.

4. Every element of the set has an inverse C^{-1} which is an element of set $CC^{-1} = C^{-1}C = E$.

The set of symmetry operations belonging to a molecule exhibits the properties defined above and as a result the methods of group theory can be used to determine the symmetry of a molecule. These properties, expressed in the form of equations, are applied to the combinations of symmetry operations.

The elements of symmetry characteristic of the geometric shape of a molecule determine the set of symmetry operations which form a separate point group denoted by a given symbol. One point group may describe the symmetry properties of a range of different molecules with the same basic symmetry. The most important point groups are listed in Table 2.1.

The point groups are donoted by the letters C, S, D, T, O with numerical and literal subscripts, e.g. C_2, $C_{\infty v}$, C_{3h}, S_2, D_{4d}, T_d, O_h.

The point groups denoted by C with a numerical subscript n contain only a rotation axis of multiplicity n, or a centre of symmetry if $n = 1$. If the letter C is provided additionally with the letter v or h in subscript, the given point group has, besides the n-fold axis, n planes of symmetry that include that axis.

Groups denoted by S with a numerical subscript, which is always even, can have an axis of rotation, a rotation–inversion axis and the centre of symmetry.

Letter D denotes point groups containing mutually perpendicular axes of rotation. The numerical subscript n denotes groups that have an n-fold axis and n twofold C_2 axes perpendicular to it and making equal angles with each other. If apart from the number n a letter is added in the subscript, the point group has also, besides the axes of rotation, planes of symmetry. When the letter d is used, these planes are diagonal and lie between the twofold axes, while the letter h means that there are n vertical

Table 2.1—Some point groups and the related symmetry elements (the numbers in the table refer to the multiplicity of the elements of symmetry)

Point group \ Element of symmetry	E	i	C_2	C_3	C_4	C_5	C_6	C_∞	S_2	S_4	S_6	σ_v	σ_h	σ_d	Example
C_1	1														$CHFClBr$, $CH_3CHClBr$
C_2	1		1												H_2O_2
C_i	1	1													$CH_3CHClCHClCH_3$ (antiperiplanar configuration)
S_4	1		1							1					
C_{1v}	1											1			$NOCl$
C_{2v}	1		1									2			H_2O
C_{3v}	1			1								3			NH_3, $HCCl_3$
$C_{\infty v}$	1							1				∞			HCN
C_{3h}	1			1									1		H_3BO_3
D_{2d}	1		3							1				2	B_2Cl_4
D_{3d}	1	1	3	1							1			3	C_6H_{12}, C_2H_6
D_{2h}	1	1	3										3σ		C_2H_4
D_{3h}	1		3	1								3	1		BCl_3
D_{4h}	1	1	4		2				1			4	1		C_4H_8
D_{6h}	1	1	6				1					6	1		C_6H_6
$D_{\infty h}$	1	1	∞					1				∞	1		CO_2
T_d	1		3	4						3			6σ		CH_4
O_h	1	1	6	4	3					3			9σ		SF_6

planes that include the n-fold axis, and one horizontal plane of symmetry perpendicular to that axis.

The point groups denoted by T describe the symmetry of tetrahedral molecules, and those denoted by O the symmetry of octahedral molecules.

2.4 SYMMETRY OPERATIONS, TYPES OF MOLECULAR VIBRATION, CHARACTER TABLES

In molecular spectroscopy, we are interested in vibrations of atoms and groups of atoms as well as vibrations of the whole molecule. The vibrations to which the molecule is susceptible are closely related to its spatial structure. The symmetry elements contained in the spatial model of the given molecule (Fig. 2.1), and the related symmetry operations, determine all the vibrational motions of the system of atoms involved.

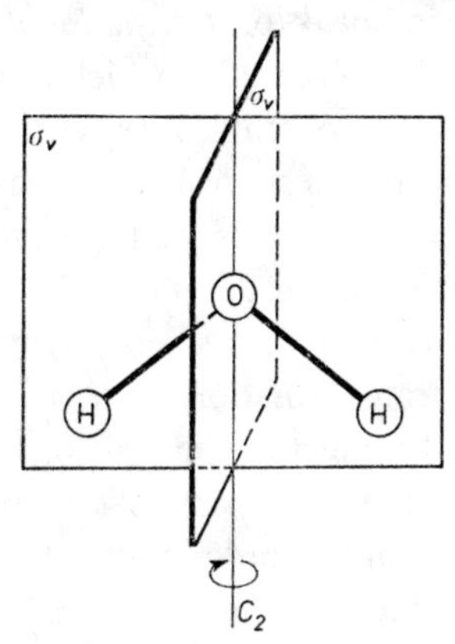

Fig. 2.1—Symmetry elements (two-fold axis C_2 and two σ_v planes) of the geometric model of the H_2O molecule belonging to the C_{2v} point group.

The symmetry operations allowed for the given molecule can be presented in a diagrammatic form, illustrating the directions of displacement of atoms under the action of particular symmetry operations. In the case of more complicated molecules this type of presentation becomes unsatisfactory. The symmetry operations can be presented more easily in terms of matrix calculus, which is an algebraic reflection of the effect of symmetry operations on the displacement vectors (Fig. 2.2). This procedure is known as the *mathematical transformation of displacement vectors.*

We consider the displacement vectors in the translational (T) and rotational (R) movement of the point representing the molecule. It is

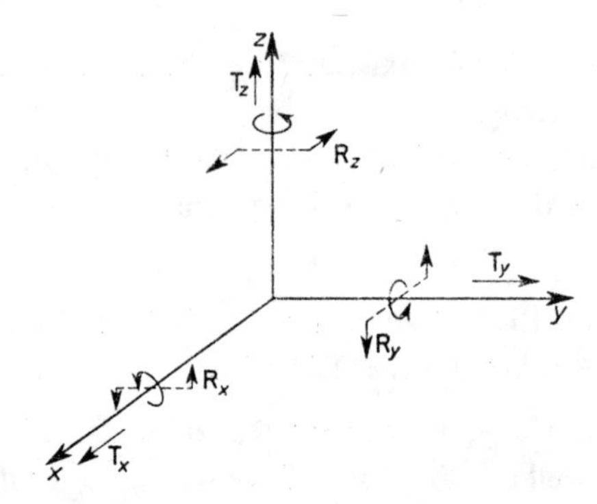

Fig. 2.2—Displacement vectors expressed in the Cartesian coordinate system: translational vectors $\mathbf{T}_x$, $\mathbf{T}_y$, $\mathbf{T}_z$ and rotational vectors $\mathbf{R}_x$, $\mathbf{R}_y$, $\mathbf{R}_z$.

most convenient to express the position of that point in the Cartesian coordinate system. We then speak of *translational vectors* $\mathbf{T}_x, \mathbf{T}_y, \mathbf{T}_z$ and *rotational vectors* $\mathbf{R}_x, \mathbf{R}_y, \mathbf{R}_z$ (Fig. 2.2) which transform independently according to a given symmetry operation represented by a transformation matrix. Such an approach allows us to treat separately the translational, rotational and vibrational movements and thus to identify the number and kind of vibrations appearing in the molecule considered.

A point group usually includes several symmetry operations, with the corresponding set of transformation matrices, called a *representation of the group*. Considering the effect of the transformation on a point representing the molecule, we obtain the simplest representation of the point group. Representations of higher order are obtained when the displacements of all the atoms in the molecule are considered independently (see Section 2.6).

The set of matrices representing the point group is given in tabular form, by summarizing in so-called *character tables*, which have the form shown in Table 2.2. To make the description of the table easier the columns have been numbered from I to IV. In the upper left corner of the table we put the symbol of the point group considered. Column II is divided according to the symmetry elements included in the given point group (in abbreviated form they are grouped in classes).

Column I gives the letter symbols, *A*, *B*, *E* and *F*, of the representation types (or more precisely the types of irreducible representations, as explained further in the text).

The particular *types of representation* correspond to various kinds of vibrations of the point representing the molecule. Those vibrations

Table 2.2—Arrangement of the character table

I	II	III	IV
Symbol of the point group	(a) Symmetry elements included in the point group or in abbreviated form: (b) Symmetry classes; the number preceding the symbol gives the number of symmetry elements contained in the class	Transformational activity	
Types of irreducible representations (vibration types)	Characters of irreducible representations assigned to symmetry elements (classes) and representation types	Rotations and translations (Cartesian coordinates) assigned to the representation types	Products of Cartesian coordinates and their combinations

are described by a set of symmetry operations included in the given point group. In spectroscopic terms they are called *types of vibration.* Vibrations are classified into types A and B depending on whether they are *symmetric* (A) or *antisymmetric* (B) with respect to the symmetry axis of greatest multiplicity. The vibrations belonging to types A and B are non-degenerate. *Doubly degenerate* vibrations belong to type E and *triply degenerate* to type F (instead of F, T is sometimes also used). Subscripts to the letters A, B, E and F introduce a further distinction within the representation types.

The numbers in column II of the table are the *characters of representation,* i.e. quantities which, when combined with the symmetry elements and types of vibration, determine precisely the transformation properties of the given point group. These characters are obtained with the aid of matrix calculus by reducing the transformation matrix to a form that cannot be further reduced. The matrices subjected to reduction are called *reducible matrices,* while those obtained after reduction are called *irreducible matrices.* Usually the initial transformation matrices, representing a given point group, are reducible, so they should be reduced. The characters given in the character table are numbers found by summation of the diagonal elements of irreducible matrices obtained by reduction. Every irreducible matrix and its character can be assigned to a suitable symmetry element (according to the columns) and to the type of representation or vibration type (according to the rows). The signs + and − appearing with the characters indicate that the vibration is symmetric or antisymmetric with respect to the given symmetry element.

In column III of the character table the symbols x, y, z (or T_x, T_y, T_z) and R_x, R_y, R_z are listed. They are assigned to various representation types. These symbols represent the translations (given in the x, y, z coordinate system) and the rotations around axes corresponding to the subscripts to R. The assignment of the translations and rotations to suitable representation types is performed on the basis of group theory independently for every point group. This means that the translations (coordinates x, y, z) and rotations (R_x, R_y, R_z) undergo transformation according to the representation type they are assigned to and belong to the vibration type associated with that representation type.

As will be shown in following chapters, a knowledge of the transformation properties of translations and rotations (coordinates x, y, z and R_x, R_y, R_z) is necessary for determining the number and types of vibrations of the molecule and for deriving selection rules for bands in infrared absorption spectra.

Column IV of the table of characters lists the products of the Cartesian coordinates x^2, y^2, z^2, xy, yz, zx, or their combinations, e.g. x^2+y^2, x^2-y^2. As in the case of translations and rotations in column III, these are also assigned to the respective representation types according to which they undergo transformation. A knowledge of the transformation properties of these functions is indispensable for determining the selection rules for bands in Raman scattering spectra.

Table 2.3 is an example of a character table, for the point group C_{3v}, presented in full and abbreviated forms; in the latter case the symmetry elements are grouped in *classes*.

Within a given point group, represented by the character table, those symmetry elements that have the same characters will belong to a common class. In group C_{3v} the symmetry elements C_3 and C_3^2 (rotation by an angle $2\times 2\pi/3$) have the same characters, so they belong to a common class. Similarly, the symmetry elements σ, σ_v' and σ_v'' belong to another common class. Grouping the symmetry elements into classes, we can write the table of characters in an abbreviated form, prefixing the symbol in the column heading with the number of equal elements of symmetry contained in that class.

In this section the character table has been discussed as a table grouping the principal parameters determining the point group. Such tables,

Table 2.3—Character table of point group C_{3v}

C_{3v}	E	C_3	C_3^2	σ	σ_v'	σ_v''	Rotations and translations	Products of Cartesian coordinates
A_1	1	1	1	1	1	1	z	x^2+y^2, z^2
A_2	1	1	1	−1	−1	−1	R_z	
E	2	−1	−1	0	0	0	(R_x, R_y), (x, y)	(x^2-y^2, xy), (yz, xz)

The same character table written in abbreviated form

C_{3v}	E	$2C_3$	$3\sigma_v$	Rotations and translations	Products of Cartesian coordinates
A_1	1	1	1	z	x^2+y^2, z^2
A_2	1	1	−1	R_z	
E	2	−1	0	(R_x, R_y), (x, y)	(x^2-y^2, xy) (yz, xz)

summarized for point groups describing molecular vibrations, can be found in appropriate textbooks [1, 4–6].

In the following section it will be shown how the selection rules for bands in Raman and infrared absorption spectra of the given molecule can be formulated.

2.5 SELECTION RULES FOR VIBRATIONAL TRANSITIONS

Absorption of infrared radiation excites the molecular vibrations which are accompanied by changes of the dipole moment $\boldsymbol{\mu}$. The dipole moment is a vector, and in the course of absorption of radiation its components (μ_x, μ_y, μ_z) are transformed as a result of the symmetry operation in the same way as translation vectors $(\mathbf{T}_x, \mathbf{T}_y, \mathbf{T}_z)$ or Cartesian coordinates (x, y, z). The changes of the dipole moment are related to the basic transitions between vibrational energy levels of the molecule on absorption of infrared radiation. The condition for these transitions to take place can be formulated as follows.

The fundamental vibrational transition is active in the infrared, if the related normal vibration belongs to the same representation as one of the Cartesian coordinates.

The types of vibrations that may be active in the infrared can thus be read directly from the penultimate column of the character table of the given point group; they are assigned to the coordinates x, y, z.

The vibration of a molecule is active in Raman scattering if it is related to a change in dipole moment due to a variation of the molecular polarizability. Polarizability is a tensor, the components of which, α_{xx}, α_{xy}, etc. are expressed by a function dependent on the products of Cartesian coordinates, x^2, xy, etc. Thus, the components of the polarizability tensor will undergo transformation as in the case of the products of the Cartesian coordinates placed as subscripts to α, or their combinations. The condition for these fundamental transitions to occur as a result of Raman scattering is as follows.

The fundamental vibrational transition is active in the Raman spectrum if the related normal vibration belongs to the same representation as one or several products of the Cartesian coordinates.

The types of vibrations that are active in Raman scattering can be read directly from the last column of the character table for the given point group; they are assigned to products of Cartesian coordinates or their combinations.

On the basis of group theory we can also determine *combination* and *overtone bands* occurring in absorption and Raman spectra. Generally speaking, the characters of these bands are equal to the products of the characters of the fundamental bands from which they are formed.

The considerations based on group theory lead to the following general rules, given without further proof.

1. A molecule that has a centre of symmetry obeys the rule of mutual exclusion, which means that a vibration that is active in infrared absorption is inactive in Raman scattering, while a vibration that is active in Raman scattering is inactive in infrared absorption.

2. A vibration which is fully symmetric is always active in Raman scattering, irrespective of the symmetry group of the molecule.

2.6 DETERMINATION OF THE NUMBER AND TYPES OF MOLECULAR VIBRATIONS

If we require information about the number of vibrations occurring in a given molecule and wish to determine, on the basis of group theory, which bands in the infrared absorption and Raman scattering spectra correspond to the given vibrations, we proceed as follows.

First we have to estimate which symmetry elements are contained in the spatial model of the molecule, and hence, to which point group the molecule belongs. In this reasoning we can proceed according to the five-step system advanced by Cotton [1].

Then, knowing the number of atoms in the molecule we can calculate the so-called *general representation* Γ of that molecule. This representation is obtained by applying all the symmetry operations possible for the given molecule to all the atoms contained in it and not only to the one point representing the molecule, as is done when calculating the simplest representation and summarizing the character table of the point group (cf. Section 2.4). We shall limit ourselves here to a conceptual description of the procedure of calculating the general representation, without giving the relevant formulae.

Our considerations are based on the vibrations of the molecule, i.e. on the resolution of the movement of the molecule into independent displacements of every atom in the directions x, y, z of the Cartesian coordinate system. We shall have $3N$ such displacements (where N is the number of atoms in the molecule), i.e. as many as the number of degrees of freedom of all the atoms in the molecule. Vibrations defined in this way include both internal vibrations of the molecule, known as *free vibrations*, and virtual vibrations, i.e. translation, and vibrations of the molecule as a whole.

Every symmetry operation applied to all atoms in the molecule is expressed by a transformation matrix with $3N$ rows and $3N$ columns. If we count the characters of these matrices (the sum of their diagonal elements), we obtain the characters of the representation Γ generated by all the vibrations of the given molecule. In this way the general representation of a molecule is found. Table 2.4 is formed from the character table of the point group D_{3h} supplemented with the characters of the general representation for a planar four-atom molecule or ion belonging to that group (e.g. BCl_3, CO_3^{2-}).

Table 2.4—Character table of the D_{3h} group supplemented with the characters of the representation of the Cartesian translations of a four-atom molecule

D_{3h}	E	$2C_3$	$3C_2$	σ_h	$2S_3$	$3\sigma_v$	Translations and rotations	Products of Cartesian coordinates
A_1'	1	1	1	1	1	1		x^2+y^2, z^2
A_2'	1	1	-1	1	1	-1	R_z	
E'	2	-1	0	2	-1	0	(x, y)	(x^2-y^2, xy)
A_1''	1	1	1	-1	-1	-1		
A_2''	1	1	-1	-1	-1	1	z	
E''	2	-1	0	-2	1	0	(R_x, R_y)	(xz, yz)
Γ	12	0	-2	4	-2	2		

Making use of suitable formulae from group theory [1] and of data from the character tables we can assign all the movements of the molecules studied to the respective representation types (modes of vibration). In the case of a four-atom molecule belonging to the D_{3h} group (Table 2.4) we get:

$$\Gamma = A_1'+A_2'+3E'+2A_2''+E'' \tag{2.1}$$

This means that the molecule performs single A_1' and A_2' type vibrations, two A_2'' type vibrations, one doubly degenerate vibration of E'' type (equivalent to two vibrations) and three doubly degenerate E' vibrations (equivalent to six vibrations). In this way we get a total of 12 movements, i.e. as many as there can be independent displacements in a four-atom molecule (3×4).

As may be seen from the considerations above, group theory allows us to calculate the total number of vibrations divided into types. In order to obtain the number of vibrations of interest from the point of view of molecular spectroscopy, we have to deduct from the total number of vibrations the translational and rotational movements of the molecule as a whole. The translations of the whole molecule along the x, y, z axes as well as its rotations with respect to those axes are considered. These movements come to a total of six (five in the case of linear molecules). The movements of the whole molecule belong to the same representation types as the translations and rotations listed in the penultimate column of the character table.

Returning to the example of the character table (Table 2.4) we find from the symbols in the penultimate column that the translations of the molecule as a whole will belong to vibrations of E' and A_2'' types and its rotations to A_2' and E'' types. Thus we have to deduct from the total number of vibrations of a four-atom molecule given by Eq. (2.1) one vibration of type A_2'', one doubly degenerate vibration E' (equivalent to two translations), one A_2' vibration and one doubly degenerate vibration E'' (equivalent to two rotations). Thus the only representations that will remain correspond to free vibrations of the molecule:

$$\Gamma_f = A_1' + 2E' + A_2'' \tag{2.2}$$

To illustrate the applications of the selection rules we shall return once more to the character table of point group D_{3h}. It follows from that table that for a molecule belonging to the point group D_{3h} the following selection rules hold.

Vibrations of A_1' and E'' types—active only in the Raman spectrum.

Vibrations of A_2'' type—active only in the infrared absorption spectrum.

Vibrations of E' type—active in both spectra.

Vibrations of A_2' and A_1'' types—inactive.

This information is obtained directly from the character table, without the need to calculate the general representation.

The number of active transitions for the given molecule (number of bands in its spectra) can be found only after the general representation related to the displacements of all the atoms in the molecule is calculated and separated into representation types and the translational and rotational movements of the molecule as a whole are subtracted from the total number of vibrations (Section 2.4).

In the case of a four-atom molecule belonging to the D_{3h} group the representation related to the free vibrations of the molecule is given by Eq. (2.2). After applying the selection rules we get:

one A_1' vibration—active in the Raman spectrum;
one A_2'' vibration—active in the infrared absorption apectrum;
two E' (doubly degenerate) vibrations—active in both spectra.

REFERENCES

[1] F. A. Cotton, *Chemical Applications of Group Theory*, Interscience, Wiley, New York, London, 1963.

[2] G. M. Barrow, *Introduction to Molecular Spectroscopy*, McGraw-Hill, New York, 1962.

[3] N. L. Alpert, W. E. Keiser and H. A. Szymański, *IR. Theory and Practice of Infrared Spectroscopy*, Plenum Press, New York, 1970.

[4] D. S. Schonland, *Molecular Symmetry. An Introduction to Group Theory and Its Uses in Chemistry*, Van Nostrand, London, 1965.

[5] H. H. Jaffé and M. Orchin, *Symmetry in Chemistry*, Wiley, New York, 1965.

[6] J. R. Ferraro and J. S. Ziomek, *Introductory Group Theory and Its Applications to Molecular Structure*, Plenum Press, New York, 1969.

3

Instrumentation and experimental techniques

H. Barańska and A. Łabudzińska

3.1 THE LASER AS A SOURCE FOR RAMAN SPECTROSCOPY

3.1.1 Principles of laser action

The phenomenon of laser action is based on the interaction between radiation and atoms or molecules which have become excited to suitable energy levels. These interactions lead to *stimulated emission of radiation* for which the wavelength, direction of propagation, phase and plane of polarization are the same as those of the incident radiation. Thus amplification of the incident radiation occurs. Hence the name of the device LASER, which is an acronym made up of the initial letters of words that form the English definition of the phenomenon: *Light Amplification by Stimulated Emission of Radiation.*

In order to amplify rather than attenuate a beam of radiation passing through it, a medium must be in an excited state which meets prescribed conditions. These conditions may be determined from considerations of absorption, spontaneous emission and stimulated emission, the three phenomena which can occur as a result of transitions between the same two energy levels E_0 and E_1 (Fig. 3.1).

According to Einstein, the spontaneous emission from an atomic system excited to the upper energy state E_1 follows the equation:

$$N_1(t) = N_1(0)e^{-t/\tau_1}$$

where $N_1(0)$ and $N_1(t)$ are the numbers of excited atoms (molecules) at the initial moment (0) and some later time (t) and τ_1 is the lifetime of the excited state. The quantity τ_1 is a measure of the duration of the excited state.

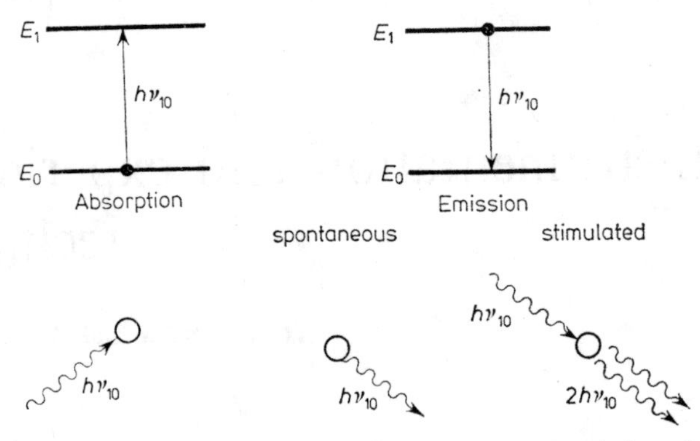

Fig. 3.1—Absorption, spontaneous emission and stimulated emission of radiation.

If an atomic system is exposed to radiation with a given frequency ν_{10} (such that $h\nu_{10} = E_1 - E_0$) then absorption or stimulated emission will occur, depending on whether the photon interacts with an atom in the ground state E_0 or in an excited state E_1. The probabilities of both phenomena taking place will then depend on the population densities of thc energy levels E_0 and E_1. If the occupation of the lower energy level is less than that of the upper level, so that $N_1 > N_0$, then stimulated emission will predominate, resulting in amplification of the incident radiation. This is due to the fact that a quantum of radiation $h\nu_{10}$, incident on an atom in the excited state E_1, causes that atom to drop to the ground level E_0; this transition is accompanied by the emission of a quantum of radiation identical to that of the incident radiation. In the absence of external radiation, an excited atom will also, after some time, drop to the ground state, owing to spontaneous emission of radiation (Fig. 3.1).

The longer the lifetime of an excited state (i.e. the longer the intervals between acts of spontaneous emission), the easier it becomes to obtain an excess population, creating suitable conditions for the stimulated emission of radiation.

In a thermal equilibrium, the distribution of atoms in different energy levels follows Boltzmann's law; that is,

$$\frac{N_i}{N_j} \approx \exp\left(-\frac{E_i - E_j}{kT}\right)$$

where N_i and N_j are the numbers of atoms in the energy states $E_i > E_j$, k is Boltzmann's constant and T is the absolute temperature. Accordingly, the largest number of atoms are in the ground state, and the population density of upper levels decreases exponentially with increasing energy level. Such a distribution is conducive to absorption of radiation at the frequency ν_{ij} (when $h\nu_{ij} = E_i - E_j$).

In order to establish conditions favourable for stimulated emission of radiation at frequency ν_{ij}, a *population inversion* must be established such that the occupation of the upper level will exceed that of the lower level. This is illustrated diagrammatically in Fig. 3.2. A system, which

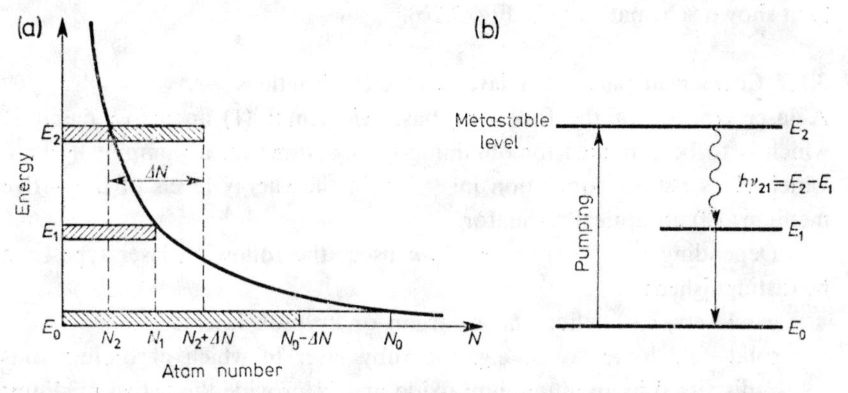

Fig. 3.2—(a) Population of energy levels in thermal equilibrium (exponential curve) and in a state of population inversion. (b) Schematic energy transitions of laser action.

according to Boltzmann's law had initially an exponential distribution of the population density of the energy levels, is transferred into an excited state by moving N atoms from the ground state E_0 to an upper excited state E_2. In this way a population inversion ($N_2 + \Delta N > N_1$) is established between the two excited levels E_1 and E_2. This type of population inversion is easier to obtain (Fig. 3.2) than a population inversion between the ground level and any of the upper excited levels. An excited system relaxes to its ground state with the emission of energy. If a radiative transition is possible between the levels having an inverted population ($h\nu_{21} = E_2 - E_1$), then stimulated emission of radiation at frequency ν_{21} can be induced by placing the active medium between two mirrors, which form a so-called *resonator*. The direct and amplified radiation (laser beam) is initiated by quanta of energy $h\nu_{21}$. These quanta, being spontaneously emitted, are trapped between the two mirrors of the resonator when passing along the optical axis of the laser. The stimulated emission then develops as an avalanche.

In order for *laser action* to be continued, an external energy supply is required to keep the system in the excited state. In other words we may say that the atoms or molecules should be continuously "pumped up" to the upper energy level, so that the population inversion will be maintained despite the steady stimulated emission of radiation. The upper excited state under discussion will have a long lifetime.

A typical scheme for laser systems is shown in Fig. 3.2b: the levels used for pumping are not the same as those in the laser transitions. A laser transition is usually accompanied by a number of other transitions (more than shown schematically in Fig. 3.2b).

3.1.2 Component parts of a laser and their functions

A laser consists of the following basic elements: (1) an active medium, which is to be activated for obtaining laser action; (2) a pumping system, which gives rise to population inversion in the energy levels of the active medium; (3) an optical resonator.

Depending on the *active medium* used, the following laser types can be distinguished:

gas lasers, e.g. helium–neon, argon or krypton lasers;
solid-state ionic lasers, e.g. the ruby laser in which chromium ions dispersed in an aluminium oxide matrix provide the active medium;
semiconductor lasers;
organic lasers;
dye lasers.

The most widely used light-sources for laser Raman spectroscopy are gas lasers such as: helium–neon (He–Ne), argon (Ar^+), krypton (Kr^+), and those using argon-krypton mixtures. These lasers emit radiation at different frequencies in the visible region (Table 3.1). Their power ranges from a few to several hundred milliwatts, and this is adequate for excitation and recording of the Raman spectrum.

The most commonly used *pumping methods*, i.e. the modes of energy transfer causing population inversions, are:

optical pumping, which consists in exposing the active medium to

Table 3.1—Wavelength and power output of laser beams emitted by the four Spectra-Physics laser types

Wavelength, nm	Laser beam power, mW			
	Krypton Model 165-01	Argon Model 165-03	Argon/Krypton Model 165-02	Helium–Neon Model 125 A
3391	—	—	—	+*
1151	—	—	—	+*
1084	—	—	—	+*
799.3	30*	—	—	—
793.1	10*	—	—	—
752.5	100*	—	—	—
676.4	120	—	20	—
647.1	500	—	200	—
632.8	—	—	—	>50
611.8	—	—	—	+*
568.2	150	—	80	—
530.9	200	—	80	—
520.8	70	—	20	—
514.5	—	1400	200	—
501.7	—	250	20	—
496.5	—	400	50	—
488.0	—	1300	200	—
482.5	30	—	10	—
476.5	—	500	60	—
476.2	50	—	—	—
472.7	—	150	—	—
465.8	—	100	—	—
457.9	—	250	20	—
454.5	—	100	—	—
351.1+363.8	—	20*	—	—
350.7+356.4	40*	—	—	—

* Laser action can be obtained by using additional optical equipment.

strong external radiation which on absorption, excites the medium to an upper energy state;

electronic pumping, employed generally in gas lasers, consisting in the excitation of gas atoms or molecules through collisions with electrons or ions during electrical discharge;

pumping, in the course of which atoms or molecules of the active medium are excited by means of energy which has been transferred from other atoms or molecules, previously excited to the upper energy state in a separate and independent process; this type of energy transfer can take place only if the energy levels of both the excited and exciting media are sufficiently close, and the transfer of energy occurs through collisions which are termed collisions of the second kind;

pumping through chemical reactions which result in the excitation of molecules.

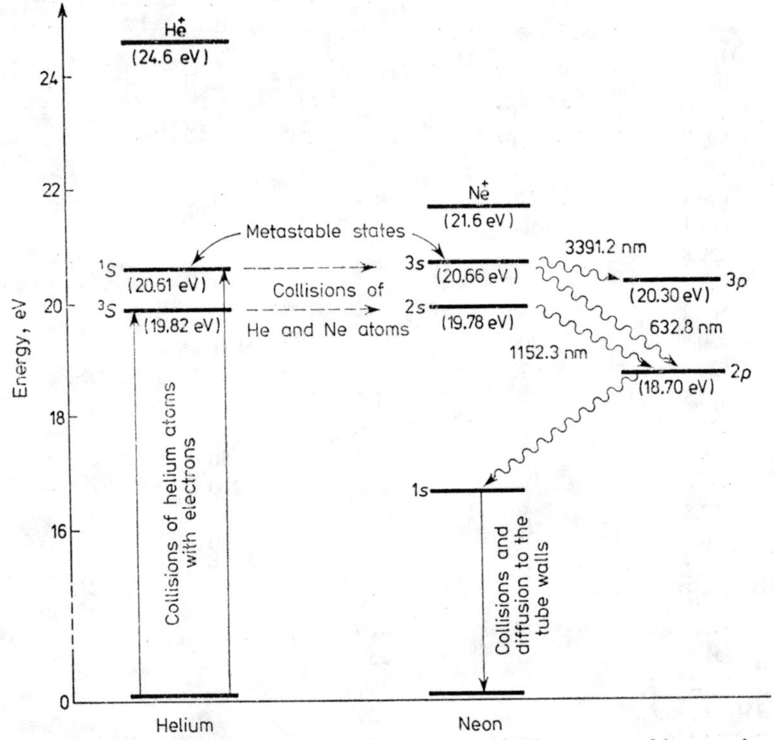

Fig. 3.3—Energy levels of helium and neon atoms in the course of laser action. Wavy lines show radiative transitions; the numbers beside them give the wavelengths of the laser beams, in nm. The levels 3*s*, 2*s*, 3*p* and 2*p* are multiple levels, between which more radiative transitions are possible than are actually shown in the diagram.

Figure 3.3 shows the energy levels of helium and neon atoms, the pumping mechanism and the transitions taking place in the course of laser action. The He–Ne laser tube is filled with a mixture of these gases. To enable pumping to the metastable energy levels to take place, as shown in the diagram, the number of helium atoms must exceed that of neon atoms tenfold (the helium pressure is approximately 100 Pa). Helium atoms are excited by collisions with electrons in the course of electrical discharge in the laser tube. The excited helium atoms transfer their energy to neon atoms during collisions of the second kind. The helium atoms in a helium–neon laser merely act as intermediates in the process of exciting the neon atoms which, being the active atoms in the system, can suffer a transition to the ground state with emission of part of their energy in the form of laser radiation. As shown in the diagram of the energy levels, laser transitions can be obtained for two lines in the near infrared region of the spectrum and one red line. In the most widely used helium–neon laser, the laser action is obtained only for the 632.8 nm red line. After falling to the lower excited state, the neon atoms can return to the ground state, releasing the excess of energy in the form of radiation or kinetic energy.

The final basic element of a laser to be discussed is the optical resonator, which includes a *chamber for the active medium*. This chamber is important because it is here that the stimulated emission takes place. In gas lasers a narrow quartz tube with flat Brewster-angle windows at each end (giving polarization of the laser beam) is usually employed (Fig. 3.4). An electric discharge is passed between two electrodes sealed into the tube, to provide electronic pumping.

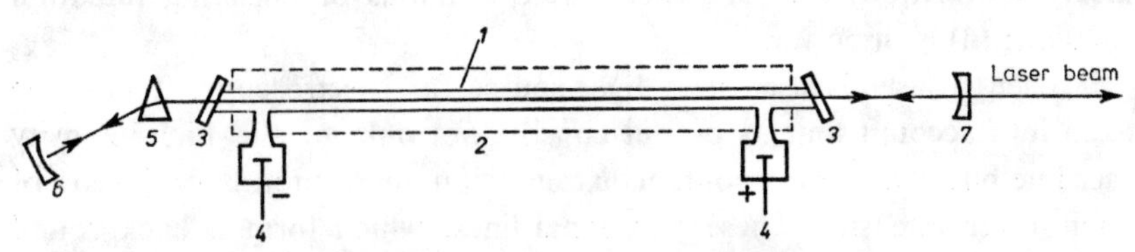

Fig. 3.4—Schematic diagram of an argon laser. (*1*) Quartz tube filled with argon. (*2*) Casing containing a solenoid (giving a magnetic field), and a cooling jacket. (*3*) Brewster-angle windows. (*4*) Electrodes. (*5*) Prism for setting laser action wavelengths. (*6*) Totally-reflecting resonator mirror. (*7*) Semi-transparent resonator mirror (about 90% reflection).

The *optical resonator* has the two functions: to establish the conditions for a high degree of stimulated emission to take place in the active medium, and to produce a monochromatic beam with a small angle of divergence.

The Fabry–Perot interferometer has been used as a resonator for some lasers. The resonator contains the active material, a rod in the case of solid lasers, or a gas-filled tube in the case of gas lasers. The resonator also has a reflecting mirror at each end of the chamber containing the active medium. These mirrors may be flat and adhere closely to the ends of a rod (as in solid lasers), or they may be combinations of two concave (Fig. 3.4), a concave and a convex, or a concave and a flat mirror. The distance between the mirrors corresponds to a multiple of the wavelength of the laser radiation, so that standing waves are set up in the active medium. This condition ensures that the photons initiating laser action travel along the optical axis of the system and are trapped between the mirrors of the resonator. In order to allow the resulting radiation beam to escape from the laser, one of the two mirrors must be semitransparent. Mirrors with low transparency, between 1 and 10%, are used, so that the laser beam is less intense outside the resonator than inside, a necessary sacrifice to maintain laser action. Emission lines from gases have measurable widths, usually of the order of 0.01 nm. The widths of laser lines are 0.01 nm or less, depending on the spectral properties of the laser medium and the optical specification of the Fabry–Perot resonator.

3.1.3 Main characteristics of the laser beam

The characteristics of a laser beam result from the particular nature of the process of stimulated emission and from the conditions under which a laser beam is generated. These characteristics are: (1) narrow spectral line-width; (2) small cross-sectional area and low divergence; (3) good linear polarization; (4) coherence; (5) continuous or pulsating radiation emission; (6) high power.

When a laser is used as a light-source for spectroscopy, it must be taken into account that its output consists not only of the high-intensity laser line but also some spontaneous emission lines (non-laser lines), of much lower intensity. These additional lines, which form a background to the laser line, should be excluded from a laser beam used as the excitation source for Raman scattering. For this purpose the beam is passed through an interference filter or a special prism block.

3.1.4 Types of laser

Lasers designed nowadays can emit beams at one or more than one wavelength. An argon laser, for example, emits beams at nine wavelengths in the visible region. An additional prism inside the resonator provides a means of selecting the wavelength of the laser beam required (Fig. 3.4). Today, the range of radiation emitted by different types of laser includes hundreds of lines from the far infrared to the far ultraviolet. In addition, there are the so-called *tunable dye lasers*, which, with a suitable choice of the dye, are capable of producing radiation of any desired wavelength in the visible spectrum. The choice of dye depends on the wavelength range of its fluorescence band, because laser action can be induced at any wavelength inside that band.

Lasers available on the market differ both in power and operating mode. There are continuously operating lasers, i.e. lasers emitting a continuous beam of constant power, and pulse-operating lasers emitting short pulses of radiation, usually of high power. The power of continuously operating lasers ranges from a few milliwatts to several watts (up to some hundred watts for the CO_2 laser), whereas the power of a single pulse-operating laser can amount to some gigawatts.

In Raman spectroscopy continuously operating lasers emitting radiation in the visible region of the spectrum with a power of up to some hundreds of milliwatts are generally used. Other laser types are used only for special purposes.

In this chapter only basic information on the principles of laser design and operation is given, with particular reference to their employment as excitation sources in Raman spectroscopy. More detailed information may be found elsewhere [1–6].

3.2 RAMAN SPECTROMETERS

The instrument used for the study of Raman spectra is now known as a *laser Raman spectrometer*, to emphasize the excitation source used [7, 8].

It consists of the following main parts: (1) excitation source; (2) optical system for illuminating the sample and collecting the scattered radia-

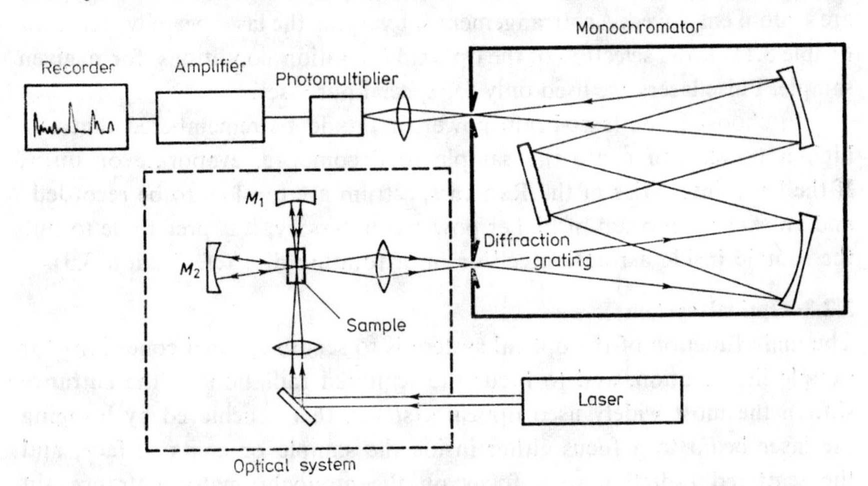

Fig. 3.5—Schematic diagram of the basic components of a laser Raman spectrometer. (M_1) Concave mirror which returns the laser beam in the direction of the sample. (M_2) Concave mirror which collimates the back-scattered radiation and reflects it into the monochromator slit.

tion; (3) monochromator; (4) detector; (5) recording device, as shown in the diagram of a simple Raman spectrometer, in Fig. 3.5.

3.2.1 Excitation source

The band intensities in a Raman spectrum may range from 0.01 to 10^{-6}% of the intensity of the excitation radiation incident on the sample. This shows that Raman scattering is far less efficient than the alternative processes of absorption or Rayleigh scattering. Therefore, powerful excitation sources and very sensitive detection systems must be used in order to obtain a Raman spectrum. Before the invention of the laser, mercury-arc lamps, emitting intense radiation at 435.8 and 546.1 nm, were used as excitation sources. Samples of relatively large (some ml) volume were required when using excitation sources of this type because the increase in the number of scattering molecules resulted in an increased intensity of scattered radiation. Spectra of gases and solids were seldom examined because their spectra were much more difficult to obtain under these conditions than were the spectra of liquids.

In modern Raman spectrometers only *laser light sources* are used (cf. Section 3.1). Their high power allows the production of spectra of an intensity sufficient for recording, even if the sample is illuminated by a point source. Helium–neon or argon lasers are normally used as excitation sources; krypton lasers or lasers with a mixture of argon and krypton are seldom employed. An arrangement for varying the laser beam wavelength (Table 3.1) allows selection of the optimal excitation conditions for a given sample. Pulse lasers are used only for special purposes.

In choosing the laser-beam power, it should be remembered that too high a power can cause the sample to decompose, evaporate or burn. If the band intensities of the Raman spectrum are too low to be recorded, and an excitation beam of higher power is necessary, it is preferable to put the sample inside a rotating cell or on a rotating disc (cf. Section 3.3).

3.2.2 Optical system

The main function of the optical system is to secure optimal conditions for sample illumination, and to focus the scattered radiation on the entrance slit. In the most widely used optical systems, this is achieved by bringing the laser beam to a focus either inside the sample or at its surface, and the scattered radiation to a focus on the monochromator entrance slit (Fig. 3.5). The intensity of the scattered radiation can be increased by mounting two additional concave mirrors M_1 and M_2 in the optical system. The first of these mirrors (M_1) refocuses the transmitted laser beam back into the sample, while the second (M_2) directs onto the spectrometer slit that radiation scattered by the sample in the reverse direction (Fig. 3.5). Both mirrors, however, are effective only if the direction of the laser beam does not deviate on passing through the sample.

The optical system also includes a polarizing device for measuring the depolarization of Raman bands, interference filters or a prism assembly to eliminate non-laser lines (cf. Section 3.1), and supplementary equipment such as filters, for the elimination of ghosts from the scattered beam (cf. Section 3.2.3), and neutral filters or iris diaphragms to reduce the beam intensity. This reduction in intensity is necessary to protect the operator's eyes while adjusting the illumination of a sample by the laser beam. Some of these component parts will be included in the optical system only in special cases.

Figure 3.6 shows the optical train used in the Spex Ramalog Raman spectrometer.

The design of the sample area facilitates the exchange of components within it, such as different cell holders for liquid-sample cells, gas-sample

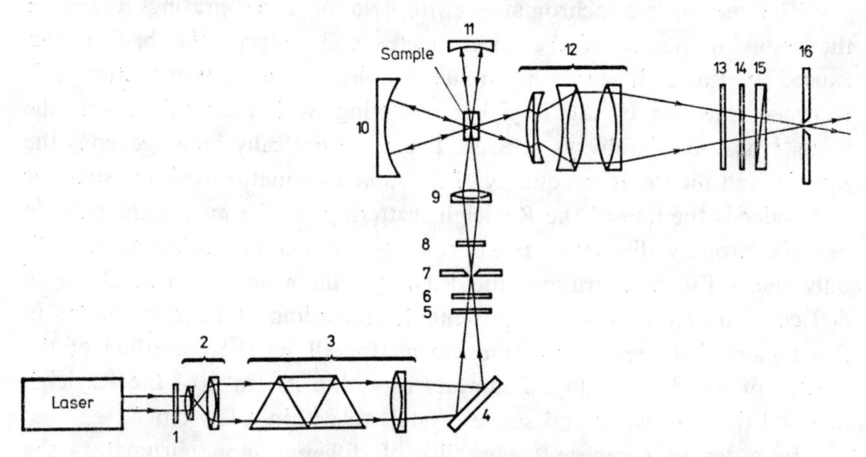

Fig. 3.6—Optical system of the Spex Ramalog spectrometer. (*1*) Optical element for rotating the plane of polarization (optional). (*2*) Optical element for widening the beam cross-section. (*3*) Prism block for eliminating the non-laser lines from the laser beam (optional). (*4*) Flat mirror. (*5*) Neutral filter. (*6*) Interference filter. (*7*) Slit (used only in combination with the prism block). (*8*) Safety filter. (*9*) Microscope objective. (*10*) Concave collimating mirror for back-scattered radiation. (*11*) Concave mirror which returns the laser beam. (*12*) Objective focusing scattered radiation onto the monochromator entrance slit. (*13*) Analyser. (*14*) Long-wave pass filter (optional). (*15*) Depolarizer. (*16*) Monochromator entrance slit.

cells, solid-sample holders, goniometers, rotating cells and discs, and cryostats. All this equipment is designed so as to permit the sample to be optimally positioned with respect to the laser beam.

3.2.3 Monochromator

As already stated, the intensity of Raman scattering by the sample is considerably lower than the Rayleigh scattering of the exciting beam. For this reason it is necessary to use a high-quality monochromator which gives a low background in the Raman spectra.

As a rule, the laser Raman spectrometer is provided with a monochromator in which the radiation is dispersed by *diffraction gratings.* A monochromator usually has two diffraction gratings, which may be mounted in the Czerny–Turner, Ebert–Fastie, or Littrow configuration [9]. Only the simplest spectrometers used for training or routine analysis are equipped with monochromators having only a single diffraction grating (Fig. 3.5). On the other hand there are research instruments available which have three-grating monochromators.

The use of monochromators with two or three gratings increases the resolving power of the spectrometer and reduces the background caused by Rayleigh scattering in the sample. When Raman scattering is recorded, the line of the Rayleigh scattering, with its maximum at the 0 cm^{-1} position, is always present. The more optically heterogeneous the sample, and the lower the quality of the monochromator used, the stronger and wider is the line of the Rayleigh scattering. In the most unfavourable case (a strongly dispersive powder sample and a monochromator with only one diffraction grating) the Rayleigh line wing can extend up to 500 cm^{-1} on either side and prevent the recording of Raman spectra in this region. A three-grating monochromator allows the recording of the spectra of powder and liquid samples to within 20 cm^{-1} of the Rayleigh line and those of gases and single crystals to within a few cm^{-1}.

In order to compare the quality of different monochromators the resolving power and Rayleigh background level close to the exciting line should be measured. If we express the Rayleigh background level as a fraction of the intensity of the exciting line, then in Coderg spectrometers, it is of the order of $10^{-11}\%$ when measured at 50 cm^{-1}, if the spectrometer is provided with a three-grating monochromator. It increases to $4 \times 10^{-9}\%$ if a two-grating monochromator is used. However, to compare background levels measured by instruments of different makes, it is necessary to have available results of measurements made under standard conditions.

If samples giving intense Rayleigh scattering are being examined, it must be made sure that no ghosts, originating from defective diffraction gratings, are also being recorded. These ghosts, which are similar in shape to Raman lines, can occur in the total spectral range recorded, since their wavelengths are equal to those of the exciting lines. The constant position of these ghosts for any given monochromator, at a particular wavelength, facilitates their identification. The intensity of ghosts increases with increase of Rayleigh scattering in the sample, and for transparent liquid samples the ghosts are practically invisible. The intensity also increases significantly with increase in the wavelength of the exciting line. Therefore it is preferable to use, as an exciting source, the blue or green lines of the argon laser, rather than the red line of the helium–neon laser. Figure 3.7 shows

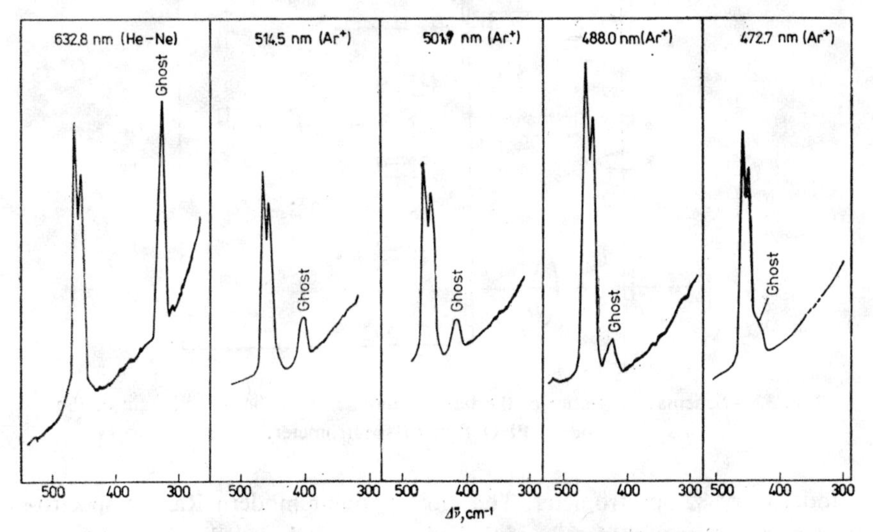

Fig. 3.7—Sections of spectra of permanent white, recorded with various exciting lines. In each section there appears a doublet, characteristic of permanent white accompanied by a ghost. The latter changes its position depending on the wavelength of the exciting line.

sections of permanent white spectra on which a ghost is superimposed. The spectra were produced with a twin-grating spectrometer equipped with replica diffraction gratings of medium quality. Today, holographic gratings of much higher quality than the replica type are generally provided.

The more diffraction gratings used in the monochromator, the lower is the intensity of the ghost lines. According to Arie [9] the ghost intensity recorded by a monochromator with only one grating is about $10^{-3}\%$

of that of the exciting line. This value drops to $10^{-8}\%$ for a twin-grating monochromator and with a three-grating monochromator ghosts are no longer measurable.

Figures 3.8 and 3.9 show the optical systems used in the Coderg Model PHO twin-grating spectrometer and the three-grating Varian

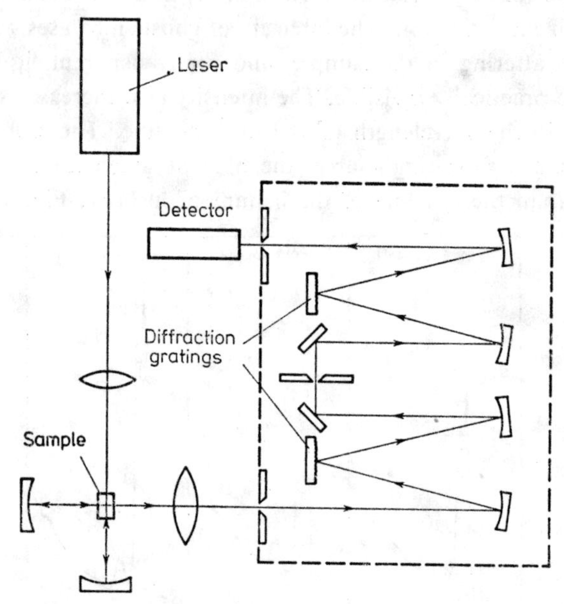

Fig. 3.8—Schematic diagram of the basic components of the optical train of the Coderg PHO Raman spectrometer.

Model Cary 82 spectrometer. The most popular modern Raman spectrometers, together with details of their detection and recording systems, are listed in Table 3.2.

A Raman spectrum is recorded as a plot of intensity of radiation scattered against frequency (wavenumber $\tilde{\nu}$). This is usually expressed as an increment $\Delta\tilde{\nu} = \tilde{\nu} - \tilde{\nu}_0$ defined as the frequency difference between the Raman band $\tilde{\nu}$ and the exciting line $\tilde{\nu}_0$. Wavenumbers can be selected by rotating the monochromator diffraction gratings, the rotation mechanism being provided with a linear wavenumber calibration.

3.2.4 Detector

A *photomultiplier* is used as the detector in Raman spectrometry. Photons incident on the photocathode cause the emission of electrons, the number emitted being proportional to that of the photons, i.e. proportional to the intensity of the incident radiation. Differences in potential between the cathode, a set of dynodes and a final anode, cause the acceleration of the photoelectrons leaving the cathode. Collisions of the electrons with

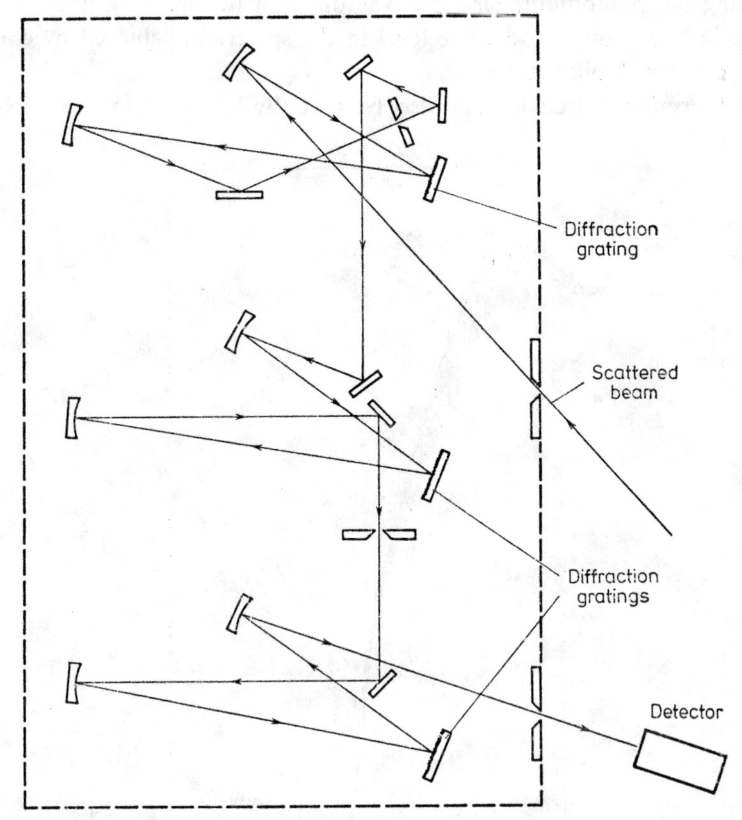

Fig. 3.9—Schematic diagram of the three-grating monochromator of the Varian Cary 82 Raman spectrometer.

dynodes, as they pass from cathode to anode, cause the ejection of secondary electrons, so amplifying the photocurrent. In a twelve-stage photomultiplier tube, the number of photoelectrons increases nearly one million times, before reaching the anode [7]. The resulting photocurrent is a measure of the radiation intensity. Unfortunately, a photocathode will also emit so-called *thermionic electrons,* which on passing to the anode produce a background for the photoelectrons. The current produced by these thermionic electrons is called the *dark current.* It augments the background level and the noise in the recorded spectrum. This is particularly disadvantageous in measuring weak Raman bands.

To reduce the number of thermionic electrons, which contribute to the current measured, spectrometers are often provided with units for cooling the photomultiplier tube. The diagram in Fig. 3.10 illustrates the drop in background and noise level in the spectrum, achieved by cooling the photomultiplier tube.

Thermionic electrons can also be partially removed from the photo-

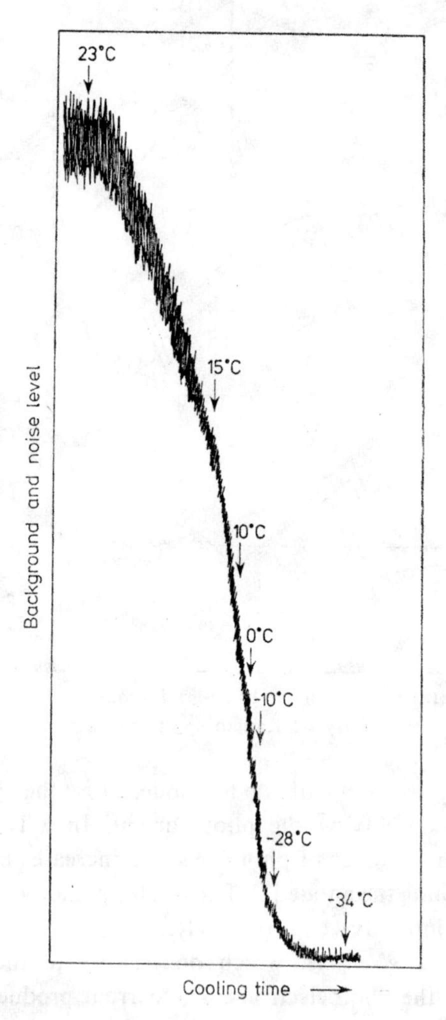

Fig. 3.10—Influence of photomultiplier cooling on background and noise levels.

electron beam by placing the photomultiplier tube in a suitably oriented magnetic field. Thermionic electrons, which leave the cathode in all directions, in contrast to photoelectrons which travel in defined paths, suffer an additional deflection in the magnetic field and are lost by collision with the photomultiplier tube walls.

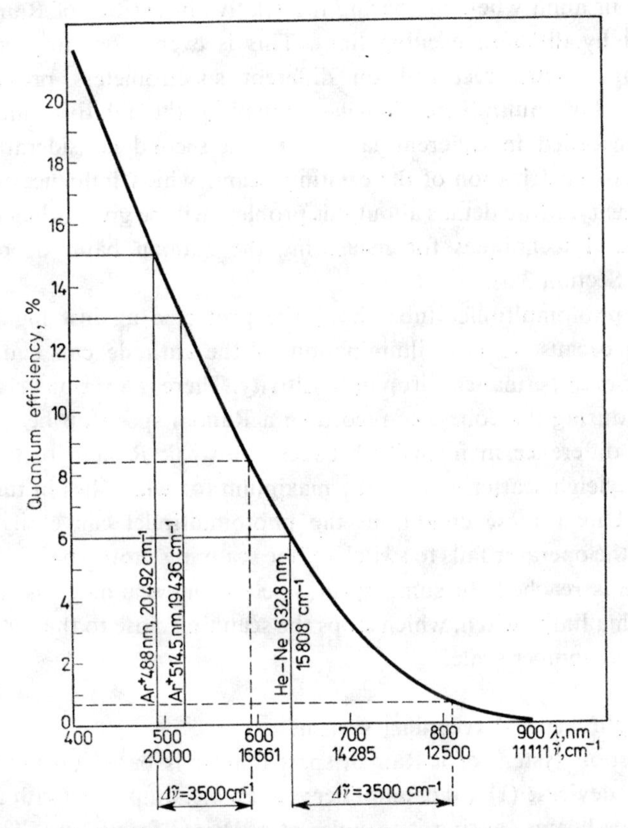

Fig. 3.11—Relationship between quantum efficiency and wavelength for the EMI photomultiplier with S-20 cathode.

Not every photon incident upon the cathode ejects an electron and the efficiency of the process for visible and infrared radiation (the regions of particular interest), depends on the photocathode response. Figure 3.11 shows the sensitivity diagram of an EMI photomultiplier provided with an S-20 cathode, which is the most widely used in Raman spectrometers. The diagram indicates the spectral ranges of Raman scattering obtained by using the 488.0 nm argon laser line and the 632.8 nm helium–neon laser line. The diagram reflects the great difference in the photomultiplier response in the two ranges and illustrates the disadvantage of using the He–Ne laser line as the exciting source, particularly for measuring the end part of the spectrum.

The variation in photomultiplier response and its non-linearity must be borne in mind when comparing the relative intensities of Raman bands produced by different exciting lines. This is even more important when comparing spectra recorded on different spectrometers provided with different photomultipliers. When comparing the relative intensities in spectra recorded in different laboratories a second consideration is the direction of polarization of the exciting beam, which influences the measured intensity. More details about this problem will be given when discussing experimental techniques for measuring the Raman band depolarization ratio (cf. Section 3.3).

The photomultiplier tube should be protected against highly intense radiation because intense illumination of the cathode can cause a temporary or even permanent drop in sensitivity. There is a serious risk of such a failure during the course of recording a Raman spectrum because of the immense difference in intensity between the weak Raman bands and the strong Rayleigh scattering line, the maximum for which lies in the position 0 cm^{-1}. Under these conditions the photomultiplier can easily be overloaded if the operator fails to switch off the scanning motor before the excitation line is reached. In some spectrometers the scanning motor is provided with a limit switch, which stops the scanning close to the zero position on the wavenumber scale.

3.2.5 Amplifying and recording systems

The detection system of a Raman spectrometer is based on the following electronic devices: (1) a d.c. amplifier; (2) an a.c. amplifier (with a chopper in the laser beam, causing it to pulse at a desired frequency); (3) photon-counting electronics (particularly useful for recording weak Raman bands). This third device amplifies each pulse of the photocurrent, resulting from the arrival of each quantum of radiation at the photocathode, and rejects pulses above or below a given height, which are due to secondary processes in the photomultiplier. The filtered signal is transmitted to the recorder.

Some spectrometers (e.g. the Cary 82, Coderg PHO) are provided with two electronic systems: a d.c. amplifier and a photon-counter, which can be used interchangeably.

Recorders are used to record spectra. Most commercial spectrometers are equipped with a drive mechanism which can operate the grating rotation and recorder-chart transport jointly or independently. Provisions are made for scanning forwards and backwards, recording two spectra (one above the other), and changing the scanning speed over a considerable range in order to contract or spread the wavenumber scale. It is possible to vary recording sensitivity by changing parameters of the electronic device selecting the power of the exciting laser beam, and adjusting the width of the monochromator slit. The wavenumber scale is printed on the chart paper by a special marker so as to mark wavenumber intervals $\Delta\tilde{\nu}$. The actual position of the monochromator gratings can be read on the wavenumber counter.

In numerous laboratories attempts have been made to couple Raman spectrometers with on-line computers or to record the spectrum on a chart suitable for computer processing (off-line). The major advantage offered by computer processing is the ability to smooth noise and record weak bands. This is realized through multiple scanning and spectral enhancement. Further advantages are automatic measurement of band position and intensity and the possibility of simultaneous determination of other quantities, such as band area, concentration of components of mixtures, and molecular constants. An additional facility is the comparison of a spectrum of a substance with spectra stored in the memory of a computer. Spex (USA) supplies an instrument called Ramacomp, which consists of a Spex Raman spectrometer coupled with a computer. This combination offers wide possibilities for measurement and processing.

3.2.6 Commercial Raman spectrometers

The performance of a Raman spectrometer depends on the number and quality of the diffraction gratings used in the monochromator, as well as on the electronics of the detector and recorder provided with the instrument. The simplest spectrometers are equipped with only a single-grating monochromator, for example the Huet R-50, or Perkin-Elmer JRS-01B; these have the lowest resolving power and the highest background level.

Much better characteristics are possessed by the twin-grating spectrometers (Table 3.2) such as the Jarrel Ash 25-300, Jeol JRS 400D, Varian Cary 83, Coderg PHO, Jobin Yvon Ramanor HG 2S and the Spex Ramalab, Ramalog and Ramacomp. The last, the Ramacomp, has the best characteristics in this class (resolving power 0.15 cm^{-1}, Rayleigh background level less than $10^{-8}\%$ at 20 cm^{-1}). The high-quality electronics and the computer provided with the instrument, together with the provision

Table 3.2—The most popular Raman spectrometers

Manufacturer	Model	Type of grating monochromator	Detection and recording	Resolution, cm^{-1}
Varian, USA	Cary 82	triple	amplified photocurrent, photon counter, recorder	0.25
Varian, USA	Cary 83	twin	amplified photocurrent, recorder	2
Spex, USA	Ramacomp	twin*	amplified photocurrent, photon counter, recorder, computer	0.15
Spex, USA	Ramalog	twin*	amplified photocurrent, photon counter, recorder	0.15
Spex, USA	Ramalab	twin	amplified photocurrent, photon counter, recorder	0.7
Jarrel-Ash, USA	400	twin	photon counter, recorder	0.25
Jarrel-Ash, USA	25-300	twin	photon counter, recorder	0.25
Jobin Yvon, France	Ramanor HG 2S	twin	photon counter, recorder, computer	0.5
Coderg†, France	PHO	twin	amplified photocurrent, photon counter, recorder	0.25
Jeol, Japan	JRS-400D	twin	photon counter, recorder	<0.6

* A third grating may be incorporated.
† This trademark went out of existence in 1976.

for fitting a third grating, allow this spectrometer to match others of the highest class. In this class is the Varian Cary 83, a three-grating spectrometer, but its detection and recording system is inferior to that of the Ramacomp.

3.3 EXPERIMENTAL TECHNIQUES

3.3.1 Main experimental difficulties and sample handling

As already mentioned in Section 1.1, the use of lasers as excitation sources has made possible the recording of Raman spectra of samples in all states of aggregation. In addition, the sample size may be reduced to a fraction of a ml or a few mg. Raman spectra of coloured samples can also be recorded if the exciting line frequency is suitably chosen, so as to avoid its absorption by the sample. For example, the He–Ne laser 632.8 nm line can be used for illumination of yellow, orange and red samples. The main difficulties in the experimental technique are caused by *fluorescence* due to sample impurities. Since there is an enormous difference in the probability and therefore the intensity of Raman scattering and fluorescence, even trace impurities can cause serious difficulties, making it impossible to measure the Raman spectrum against an intense fluorescence background. For example, traces of greases used in distillation assemblies (to lubricate ground-glass joints) or various products of oxidation, which very often occur in chemical samples, give rise to fluorescence even when the sample is illuminated with the red He–Ne laser line. Sometimes, the fluorescence of impurities can be eliminated by exposing the sample to intense white light but a high-intensity laser beam is much more effective. The strong fluorescence observed with these impurities is due to the strong absorption of the exciting beam energy. This often leads to photochemical reactions resulting in the production of non-fluorescent substances. An effective method of purification is *distillation* or *sublimation* under reduced pressure, performed in an apparatus without lubricant. Good results can also be obtained by *filtering* liquid samples or solutions through active charcoal or aluminium oxide layers in a glass tube, which is then used as the sample cell for spectrum recording. The adsorbent layer is placed between plugs of cotton or glass wool and the liquid sample is introduced above the adsorbent. The tube is then sealed and the liquid centrifuged into the space below the adsorbent. For recording the spectrum, this part of the tube is placed in the laser beam. In multicomponent samples, different chromatographic techniques may be used for component separation.

A *solvent* suitable for Raman spectrometry is water, which has a poor

Raman spectrum. Carbon disulphide and carbon tetrachloride also have poor Raman spectra and can, therefore, be used as solvents. The lower aliphatic hydrocarbons and alcohols are sometimes used as solvents because their Raman bands are weak. Before use all solvents should be distilled in a lubricant-free distillation assembly or filtered through aluminium oxide. When preparing solutions, it should be remembered that the limit of detection is of the order of 1%.

Crystalline powders can also be examined in the form of pressed pellets. Potassium bromide or powdered polyethylene is sometimes used as an additive for tablet-making.

3.3.2 Sample illumination and sample cells

The scattered radiation is commonly viewed at 90° to the direction of the exciting beam, thus practically eliminating the primary beam. The intensity of Rayleigh scattering is then about 0.1% of the exciting beam intensity, and the Raman scattering intensity is about 10^{-5}% of that. Observation of scattered radiation viewed at 180° may be useful in sample-surface examination. Such an optical system is employed in one of the Varian Raman spectrometers. Both types of illumination are shown in Fig. 3.12.

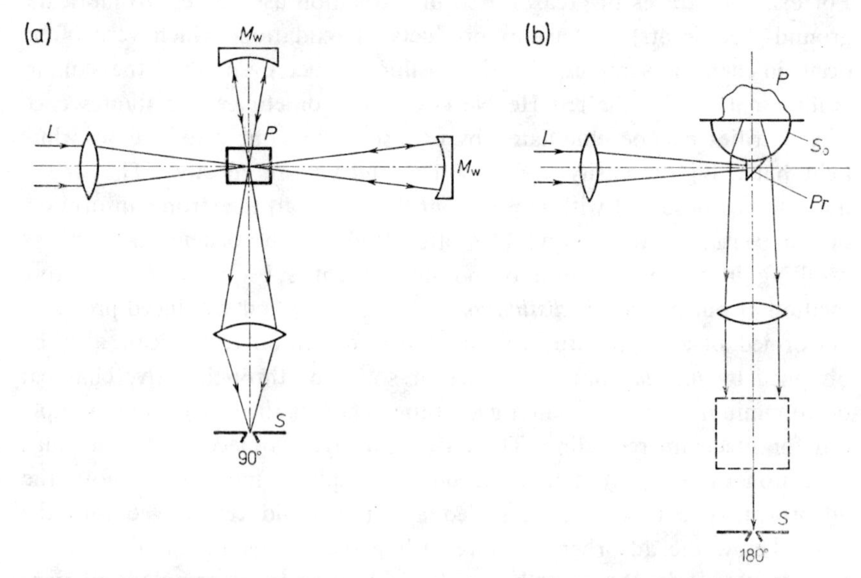

Fig. 3.12—Arrangements for Raman scattering observation at 90° (a) and 180°. (b) (*L*) Laser beam. (M_w) Concave mirror. (*P*) Sample. (*S*) Slit. (S_0) Hemispherical lens. (*Pr*) Prism.

If two concave mirrors are added to the illumination–observation system, where the sample is viewed at 90° (Fig. 3.12a), and the sample is transparent to the laser beam, the recorded scattered radiation intensity may increase fourfold from that obtainable without the mirrors. In the 180° system, a small prism directs the laser beam to the examination point within the sample. The back-scattered radiation passes through a hemispherical lens about 1 cm in diameter and is then directed by a special optical train to the entrance slit of the monochromator.

The simplest sample cell for liquids and powders is the ordinary glass capillary tube used for melting point measurements. The sample is placed in the capillary tube and the ends are sealed. This prevents the sample from suffering oxidation, and absorbing moisture and other impurities. Glass phials, ampoules or even flasks may also be used as sample cells. Transparent solid samples, e.g. polymers, in the form of a rod or fibre may be placed directly in the laser beam. Raman spectra of opaque solid samples can be obtained by surface scattering of the exciting beam. Figure 3.13 illustrates

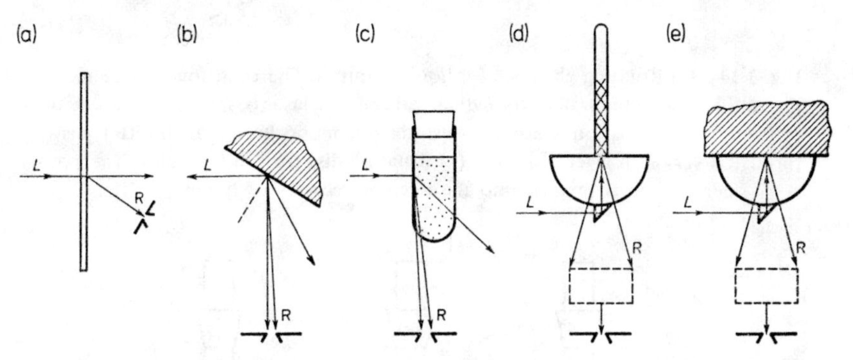

Fig. 3.13—Examples of sample positioning in the laser beam. (a) Capillary tube. (b) Rod. (c) Powder in a glass phial observed at 90°. (d) Liquid in a capillary tube. (e) Rod observed at 180°. (*L*) Laser beam. (R) Radiation scattered towards the monochromator slit.

various ways of positioning the sample in the laser beam. Coloured samples, which absorb energy from the exciting beam and may be decomposed by heat, are located in a rotating cell (liquid samples) or a rotating disc (powder samples). Kiefer and Bernstein [10] present some basic arrangements of such devices (schematic diagrams of which are given in Fig. 3.14). A small electric motor (with speed variable up to 3000 rpm) is used to drive the device. The operating speed is chosen experimentally, so as to obtain the maximum intensity of the recorded Raman bands.

For the quantitative analysis of transparent liquid samples, rectangular glass cells (Fig. 3.15) are commonly used. The cell is placed in the laser beam, in a fixed position with respect to the monochromator entrance slit, by means of a special cell holder. Proper adjustment of the exciting beam is a condition for good reproducibility of the beam intensity measurements.

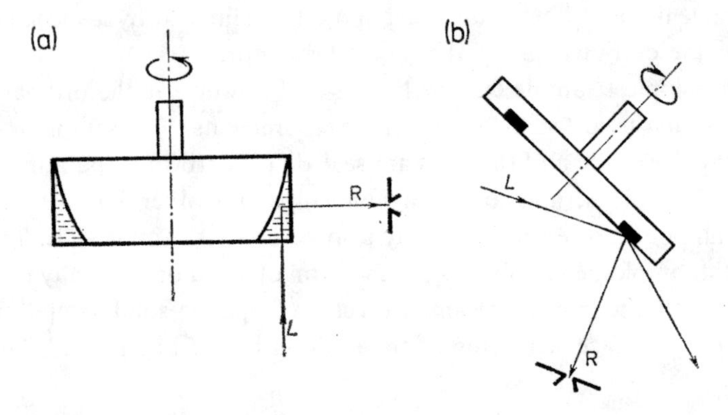

Fig. 3.14—(a) Rotating glass cell for liquid samples. The centrifugal force moves the liquid towards the wall of the cylindrical cell; the laser beam L passes along the cylinder axis; radiation scattered towards the monochromator slit (R) passes through a very thin layer of liquid. (b) Rotating disc for solid samples. The powder sample is pressed into the circular recesses of the metal disc.

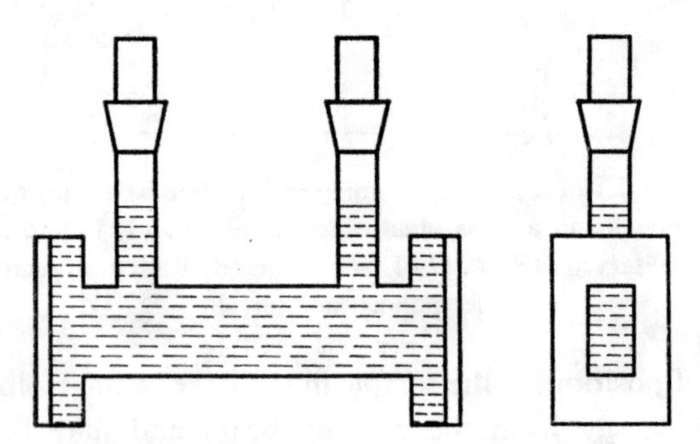

Fig. 3.15—Rectangular liquid sample cell for quantitative measurements.

Single crystals generally give good Raman spectra. A single crystal may be placed on a goniometer head, and the spectra recorded for different orientations of the optical axes to the coordinates x, y, z of the crystal [7, 11]. Different Raman spectra are obtained at different orientations of the crystal. By use of polarized light, the components of the polarizability tensor may be calculated for some classes of crystals.

Gases and vapours are examined in cells of various types, the simplest being shown in Fig. 3.16. Because in a gas cell there is a relatively small

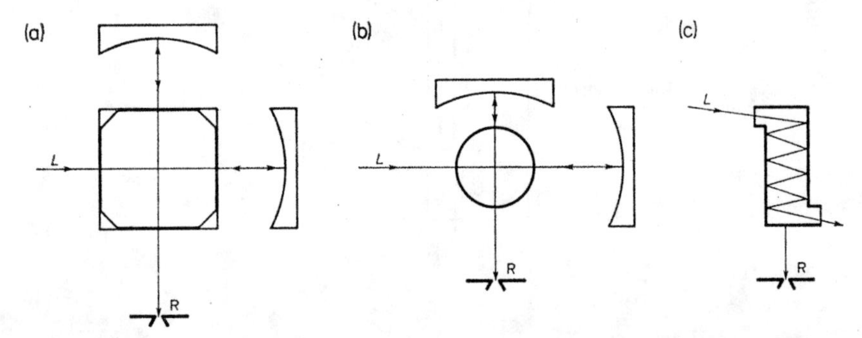

Fig. 3.16—Gas and vapour sample cells. (a) Cell with four flat parallel glass windows. (b) Glass cylinder. (c) Multipass cells. (*L*) Laser beam. (R) Radiation scattered towards the monochromator slit.

concentration of sample molecules, the path length should be increased by passing the exciting light several times through the cell, as in multiple-reflection cells (Fig. 3.16c). Vapours separated by gas chromatography can be examined in simple cylindrical glass containers.

If desired, samples can be melted (Fig. 3.17) or cooled to low temperature (Fig. 3.18).

3.3.3 Band position measurement

The basic parameter used for identification of a molecule is the band frequency. If accurate wavenumber values are required, the wavenumber scale should be calibrated. Most commercial Raman spectrometers allow the operator to set any exciting line on the zero point of the counter scale, on which the differences between the wavenumber of the measured band and that of the exciting line can be read. The setting of the exciting line can be checked by inserting an attenuation filter (e.g. sheets of blotting paper) before the entrance slit of the monochromator, and recording the exciting line on chart-paper. It should be remembered that an exciting beam of too high an intensity can damage the photomultiplier! The Rayleigh scattering of any sample is often sufficiently strong for recording the position of the exciting line. The slit-width of the monochromator and the photomultiplier voltage should be reduced, but the scanning and

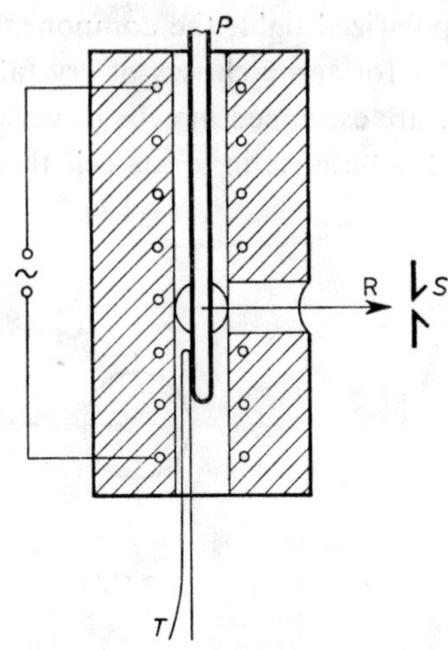

Fig. 3.17—Device for examining molten samples. (*P*) Glass cylinder with a molten sample. (R) Radiation scattered towards the monochromator slit *S* at 90° to the direction of the laser beam, which illuminates the sample in a direction perpendicular to the plane of the diagram. (*T*) Thermocouple.

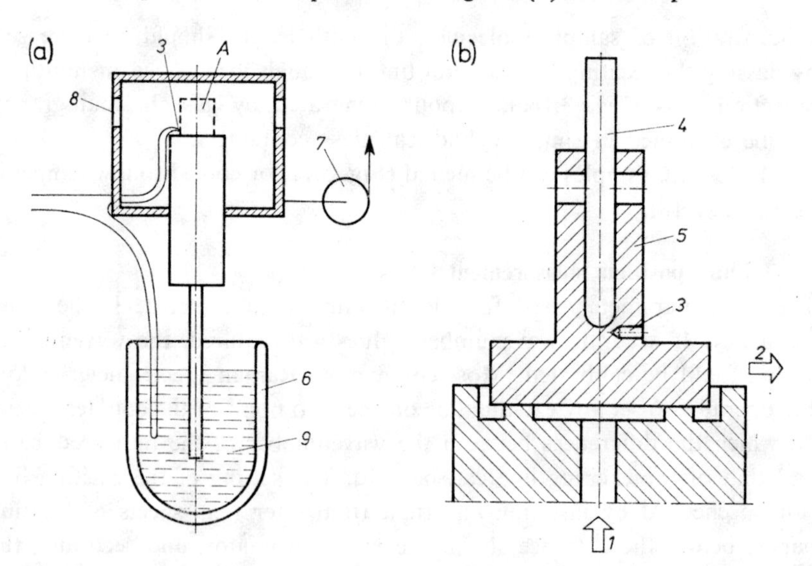

Fig. 3.18—Cryostat. (a) Schematic diagram of the arrangement. (b) Detail *A* enlarged. (*A*) Sample compartment with a sample in a sealed capillary tube. (*1*) Direction of flow of liquid nitrogen. (*2*) Direction of nitrogen gas outflow. (*3*) Place for thermocouple loop. (*4*) Capillary tube with sample. (*5*) Copper block with holes for laser and scattered beams. (*6*) Vacuum flask. (*7*) Vent for nitrogen gas. (*8*) Cryostat windows. (*9*) Liquid nitrogen.

chart speeds should remain the same as for recording the sample spectrum. An easy method of calibrating the total scale of the instrument is by recording the non-laser lines of argon (or neon) when the Ar^+ (or He–Ne) laser is used. In this case the absorbing filter, located before the sample,

Table 3.3—The stronger non-laser lines of He–Ne and Ar^+ lasers

He–Ne laser 632.8 nm excitation line		Ar^+ laser 488.0 nm excitation line	
$\Delta\tilde{\nu}$, cm^{-1}*	Relative intensity†	$\Delta\tilde{\nu}$, cm^{-1}*	Relative intensity†
6.8	m	37	s
15.6	vs	103	s
58.9	m	220	vs
93.4	m	261	m
135.6	vs	352	vs
179.9	m	380	vs
182.7	vs	528	vs
201.0	m	561	vs
230.1	m	737	vs
285.7	m	1043	m
435.0	vs	1057	s
485.0	m	1134	m
648.2	vs	1174	m
657.3	m	1578	s
769.2	m	2503	m
802.6	m	2657	m
828.1	s	3510	m
1370.8	vs	3578	m
1565.0	s		
1582.0	vs		
1635.8	m		
1862.4	vs		
1999.4	vs		
2358.8	m		
2448.5	s		
2531.5	m		
2546.1	m		
3212.0	m		
3428.9	m		
3483.9	m		
3510.9	m		

* $\Delta\tilde{\nu} = \tilde{\nu}_{ex}$ (exciting radiation) $-\tilde{\nu}_r$ (recorded radiation).
† vs—very strong line, s—strong, m—medium.

should be removed from the instrument, so that the non-laser lines which are absorbed by this filter can be recorded. Instruction manuals supplied with commercial spectrometers, and books on Raman spectroscopy, e.g. [12], include tables of the accurate frequencies of non-laser lines. A selection of these lines for neon and argon is given in Table 3.3. Spectra of substances with precisely determined band positions, e.g. indene (Table 3.4), can also be used. Recorded calibration spectra are used to correct band frequencies measured in sample spectra.

When identifying a sample, it should be remembered that spurious bands, which do not belong to the sample, may have been recorded within the spectrum. These bands can be due to various causes.

(a) Bubbling, caused by boiling, degassing, or photodecomposition. Similar effects are observed with suspensions, the particles of which migrate through the laser beam. These spurious bands are random in occurrence and can be identified immediately by repeated recording of the spectrum.

(b) The use of too strong an exciting laser beam. In instruments equipped with an interference filter rather than a prism, this filter, which absorbs the non-laser lines, becomes degraded with use and begins to pass radiation at these wavelengths. These lines can easily be identified because they appear in each spectrum of a powder sample; they can also be identified from data given in tables or from any spectrum of the non-laser lines. Non-laser lines near the exciting line often appear in the spectrum because of the band-width of the interference filter.

(c) Mechanical defects in the monochromator gratings, resulting in grating ghosts. This is particularly important for a twin-grating monochromator because a three-grating monochromator is, in general, ghost-free. Ghosts are characteristic of a given set of gratings and a given exciting laser line. Thus, ghosts always appear at the same position in the spectrum, but the ghost band profiles will vary with the sample examined. If the exciting beam is strongly scattered at the sample surface or by random reflections at the cell walls, the ghosts are very strong. Ghosts characteristic of any particular instrument can be identified when the laser beam is scattered by a metal surface or a permanent white surface.

3.3.4 Band intensity measurement

The accuracy of a quantitative band intensity measurement depends on the reproducibility of this measurement. The measured value of the Raman band intensity can be influenced by the following factors: spectral distribu-

Table 3.4—The stronger Raman bands of liquid indene

$\Delta\tilde{\nu}$, cm^{-1}	533.7	730.4	1018.3	1205.6	1552.7	1610.2	2892.2	2901.2	3054.7
Relative intensity*	s	vs	vs	vs	vs	s	s	m	m

* vs—very strong band, s—strong, m—medium.

tion of the photomultiplier response, sample temperature, absorption of radiation by the sample, depolarization ratio of the band, and optical effects. As intensity measurements are comparative, attention must be paid to factors which differ for the samples being compared or the sample and standard.

If different bands of the same Raman spectrum are being compared, attention must be paid to intensity variations due to the *photomultiplier response*. Manufacturers usually supply a graph of photomultiplier response as a function of frequency. If this information is not available, the instrument should be calibrated, e.g. by using an incandescent lamp having a known light intensity distribution as a function of wavelength.

Raman spectra are normally recorded at room temperature and a *temperature variation* of a few degrees does not influence band intensities. If, however, intensities obtained at very different temperatures are being compared, the temperature dependence of any particular band must be examined because it may differ for different bands.

The *absorption of radiation* by colourless samples may be neglected, but an attempt should be made to reduce any strong absorption by coloured substances by positioning the laser beam so that there is only a thin layer of sample between the beam and the monochromator slit.

Differently polarized components of the Raman scattering are absorbed and reflected in a different way by elements of the optical train. In Raman laser spectrometers with grating monochromators the most important factor is the dependence of the monochromator transmittance on the polarization of the transmitted light. For this reason spectrometers are provided with *depolarizers* (e.g. quarter-wave plates) placed before or behind the monochromator entrance slit.

The *optical effect* is due to samples having different indexes of refraction (Chapter 6).

Raman band intensities are best compared by means of a quantity known as the *integrated band intensity* (cf. Section 8.6). This depends on the natural width of a band and is not affected by the slit function. The natural width of a band depends mainly on the transition probability between the broadened energy levels of a molecule, and the integrated band intensity thus has a real physical significance. On the other hand, the *peak intensity* (cf. Section 8.6) depends in addition on the slit function, so that ideally, in quantitative analysis, integrated band intensities should be measured, but this is time-consuming without the aid of a computer. However, if there is no possibility of using a computer, the peak intensity has to be used instead. This will be satisfactory for comparative measurements based on the same analytical bands measured under the same experimental conditions, provided that the half-widths of the analytical bands are equal.

3.3.5 Depolarization ratio measurement

The plane polarization of laser beams makes the depolarization ratio measurement quick and easy. The depolarization ratio of opalescent liquids and powder samples cannot be measured, because of depolarization of the scattered light, caused by multiple reflections within the sample. The depolarization ratio of single crystals can be measured if the crystal is large enough to be mounted on a goniometer head.

The principle of depolarization ratio measurement is illustrated in Fig. 3.19. A quarter-wave plate is located before or behind the monochromator entrance slit. This plate is used to change linear into circular polar-

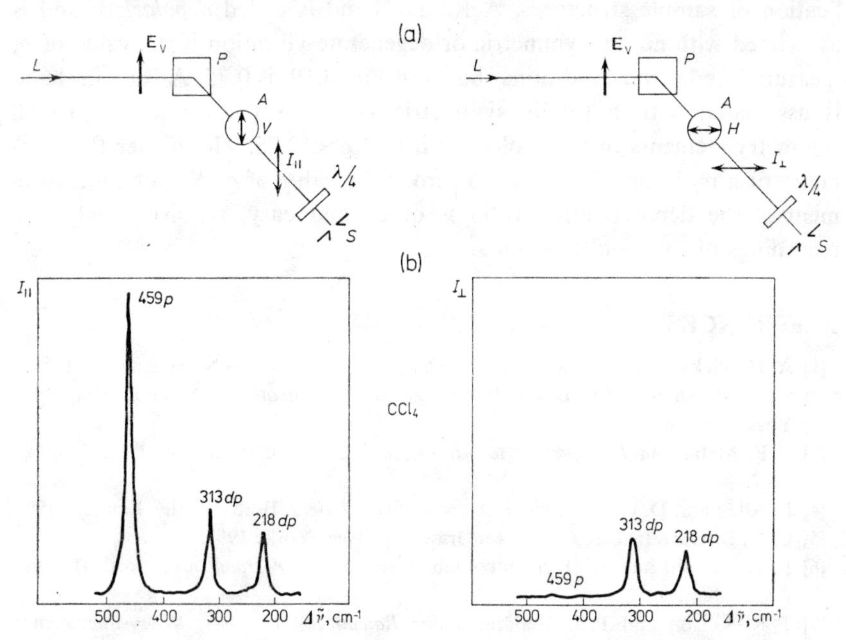

Fig. 3.19—Depolarization ratio (ϱ) measurement. (a) Schematic diagram of a measuring arrangement. (b) Spectrum of CCl_4. (*L*) Laser beam. (*P*) Sample. (*A*) Scattered beam polarization direction analyser. ($\lambda/4$) Quarter-wave plate. (*S*) Monochromator entrance slit. (**E**$_v$) Direction of laser beam electric vector (perpendicular). ($I_{\parallel}$) Intensity of the scattered-light component with its electric vector parallel to that of the laser beam. ($I_{\perp}$) Intensity of the scattered-light component with its electric vector perpendicular to that of the laser beam.

ization; this is necessary because of the variable monochromator transmittance of linearly polarized light. A grating monochromator has a higher transmittance when the polarization direction is parallel to the entrance slit than when it is perpendicular to it. To identify the polarization direction in a scattered beam, polaroid plates (marked $||$ and $\perp$, or V and H) can be used. The performance of any experimental arrangement can easily be checked by measuring the depolarization ratio of the bands of pure carbon tetrachloride.

The *depolarization ratio* is given by $\varrho = I_{\perp}/I_{||}$ where $I_{||}$ is the intensity of the scattered light component, the electric vector of which is parallel to that of the incident beam, and $I_{\perp}$ is the intensity of the scattered light component, the electric vector of which is perpendicular to that of the incident beam.

Information on band depolarization ratios is useful in the identification of sample structures. A Raman band is called *depolarized* and is associated with an antisymmetric or degenerate vibration if the value of ϱ, measured under the conditions shown in Fig. 3.19, is 0.75. A Raman band is associated with a totally symmetric vibration (with reference to all symmetry elements of the molecule) if $0 \leqslant \varrho < 0.75$. The higher the symmetry of a molecule, the closer to zero is the value of ϱ. A rough measurement of the depolarization ratio is quick and easy, requiring only two recordings of the band examined.

REFERENCES

[1] A. H. Piekara, *The New Image of Optics* (in Polish), PWN, Warszawa, 1968.
[2] S. L. Marshall (ed.), *Laser Technology and Applications*, McGraw-Hill, New York, 1968.
[3] T. P. Melia, *An Introduction to Masers and Lasers*, Chapman & Hall, London, 1967.
[4] L. Allen and D. G. C. Jones, *Principles of Gas Lasers*, Butterworths, London, 1967.
[5] C. G. B. Garrett, *Gas Lasers*, McGraw-Hill, New York, 1967.
[6] F. T. Arecchi and E. O. Schultz-Dubois (eds.), *Laser Handbook*, North-Holland, Amsterdam, 1972.
[7] T. R. Gilson and P. J. Hendra, *Laser Raman Spectroscopy*, Wiley-Interscience, London, 1970.
[8] M. C. Tobin, *Laser Raman Spectroscopy*, Wiley-Interscience, New York, 1971.
[9] B. F. Mentzen (ed.), *Spéctroscopies infrarouge et Raman*, Masson, Paris, 1974.
[10] W. Kiefer and H. J. Bernstein, *Appl. Spectrosc.*, 1971, **25**, 500, 609.
[11] S. P. S. Porto, J. A. Giordmaine and T. C. Damen, *Phys. Rev.*, 1966, **147**, 608.
[12] J. Loader, *Basic Laser Raman Spectroscopy*, Hayden/Sadtler, London, 1970.

4

Identification of organic compounds

J. Terpiński

4.1 IDENTIFICATION RULES: THE CONCEPT OF GROUP FREQUENCIES

Raman spectra, which like infrared spectra are of the vibrational–rotational type, provide in encoded manner details of the structure of the substance being examined. Thus these spectra can be used for identifying unknown substances, detecting particular atomic groups and bond types linking the atoms, defining the geometric structure of molecules, and total analysis of vibrations.

An unknown substance can be identified by simply comparing its spectrum with a library of the spectra of known compounds and selecting that spectrum having an identical appearance to the one being examined. Only molecules with the same atomic configuration have identical spectra. Without preliminary attempts to assign its bands to particular chemical groups, the Raman spectrum is seldom used for identifying unknown substances. The simple spectrum-matching method is more commonly used for identifying substances by their infrared spectra, since reference spectra can easily be found in many readily accessible catalogues, and simple and inexpensive infrared spectrometers can be found in most laboratories. The documentation of Raman spectra is, as yet, not sufficiently developed on the international scale for this purpose. However, among the few catalogues which have been compiled, the Sadtler catalogue and the DMS catalogue, collected by Schräder and Meier, comprise almost one thousand Raman spectra of organic compounds of many chemical classes. In some research centres, small-scale attempts have been made to encode Raman spectra so as to prepare them for computer-aided selection. The problem of interpreting Raman spectra and ascribing their bands to particular structural elements of the molecule examined is still of great importance. The principles of interpretation are, of course, the same as those for infrared spectra. A molecule that consists of N atoms can have $3N-6$ (or $3N-5$ if it has a linear structure) normal vibrations. When a molecule is interacting with electromagnetic radiation, it absorbs suitable energy quanta, which excite the particular normal vibrations and give rise to absorption bands which appear at the appropriate wavenumber or frequency. The frequencies of these vibrations can be calculated for simple molecules that consist of a few atoms and are of sufficiently high symmetry. Such a calculation requires information on the spatial arrangement of the atoms within the molecule and on force constants and is therefore time-consuming. The spectra of polyatomic molecules can be interpreted by extrapolation from the results obtained for simple molecules and use of the empirical knowledge acquired from the study of the spectra of a large number of more complex molecules. For this purpose the *concept of localized vibrations* is useful. If, owing to the excitation of one normal vibration of the molecule, the oscillation amplitude of one molecular fragment is much larger than those of other groups of atoms, then this vibration is spoken of as a localized vibration. This phenomenon is caused by the weak coupling of this vibration with other vibrations of the remaining atoms in the molecule. Consequently, if the remaining part of the molecule is modified, the frequency of the localized vibration will change only slightly. The more the normal vibrations of bonded oscillators differ from one another, and the greater the distance between the vibrating groups, the weaker the coupling will be. The frequency of the localized vibration is thus characteristic of the atomic grouping and always gives rise to bands in the Raman or infrared spectra at approximately the same position. The probability of excitation of a particular vibration is determined by the so-called *selection rules*, which can be derived from the application of group theory to atomic vibrations in molecules belonging to different classes of symmetry (see Chapter 2). Characteristic bands must be intense if they are to be easily identified and allow detection of the presence of certain groups in the molecule examined.

4.1.1 Band intensities

The Raman band intensity depends on the change of polarizability during vibration of a group of atoms and is proportional to the expression $(\partial\alpha/\partial Q)^2$ where α is the polarizability and Q the normal coordinate. On the other hand, the infrared band intensity depends on the dipole moment change during vibration, and is proportional to the expression $(\partial\mu/\partial Q)^2$. Thus it might be expected that some groupings will give rise to strong bands in the Raman spectrum, whereas others will give strong bands in the infrared spectrum. Thus, some groupings can be more easily identified from the Raman spectrum, while others are more readily identified from the infrared spectrum. The maximum amount of information about the structure examined can be obtained if both infrared and Raman spectra are available. Therefore, if possible, in the following discussion, characteristic bands are studied in both the Raman and infrared spectra.

Band intensities in both types of spectra can be predicted with great reliability by using the following general rules.

1. Non-polar or slightly polar groups tend to have strong Raman bands, whereas strongly polar groups have strong bands in the infrared spectrum. There are, however, some exceptions, for example the band for the C≡N group is very strong in the Raman spectrum but often weak in the infrared spectrum.
2. According to the alternative exclusion rule, molecules that have a centre of symmetry give rise to bands which appear at different wavenumbers in the Raman and infrared spectra.
3. The C—H stretching bands of aliphatic groups are strong in the Raman spectrum but weak in the infrared spectrum. The intensity of these bands is of course proportional to the C—H bond number in the given molecule. On the other hand, the C—H bending bands are of medium strength in the infrared spectrum but weak in the Raman spectrum.
4. The C—H stretching bands of vinyl groups or aromatic rings are of medium strength in the Raman spectrum and weak in the infrared spectrum.
5. The out-of-plane C—H bending vibrations of unsaturated systems (vinyl groups, aromatic compounds) have strong bands only in the infrared spectrum.
6. The C—H stretching band of acetylene is weak in the Raman spectrum and of medium intensity in the infrared spectrum.
7. The stretching bands of polar groups, such as O—H and N—H (particu-

larly the former), are strong in the infrared spectrum but weak or very weak in the Raman spectrum. In addition, the bending bands of these groups are more intense in the infrared spectrum than in the Raman spectrum.

8. The stretching vibrations of the single bonds C—C, N—N, S—S and C—S give rise to strong bands in the Raman spectrum but weak bands in the infrared spectrum.
9. The stretching vibrations of the multiple bonds C=C, C=N, N=N, C≡C and C≡N tend to give rise to very strong bands in the Raman spectrum and weak or very weak bands in the infrared spectrum. On the other hand, the C=O stretching vibration has a very strong band in the infrared spectrum. This band is of only medium strength in the Raman spectrum.
10. Raman spectra of ring compounds contain only one strong band, which is characteristic of the fully symmetric (breathing) vibration of the ring. The frequency of this vibration allows the ring size to be determined.
11. Aromatic compounds give rise to a series of sharp and strong bands in both the Raman and infrared spectra.
12. Groupings of two identical, directly coupled oscillators, for example H—C—H and C—O—C, have two stretching vibrations, a symmetric and an antisymmetric. The symmetric band is much stronger in the Raman spectrum whereas the antisymmetric band is stronger in the infrared spectrum (cf. Section 4.1.3.1).
13. Overtone and combination bands are stronger in the infrared spectrum than in the Raman spectrum, in which they are very seldom observed.

4.1.2 Spectral regions of characteristic group frequencies

The frequency is characteristic of a particular group when it differs from the frequencies of other groups situated in its near vicinity (a special case is that of directly coupled oscillators, discussed later in this chapter). The characteristic frequency of a diatomic fragment of a molecule can be roughly determined when this fragment is considered as a harmonic oscillator. The frequency of the oscillator is given by the expression $\nu = \frac{1}{2\pi c}\sqrt{f/m_r}$, where f is the force constant and m_r* is the reduced mass.

* See p. 186 for definition.

The greater the force constant, the more difficult it is to move the atoms from their equilibrium position. Therefore the stretching frequencies of double and triple bonds are higher than those of single bonds. In a polyatomic molecule the frequencies of vibrations of groups with double and triple bonds differ markedly from those associated with single bonds in the molecular skeleton and so are characteristic group frequencies. The frequency of the stretching vibrations of groups with a low m_r value is high and characteristic of these groups (e.g. the X—H group where H is the hydrogen atom and X is a heavier atom such as carbon, nitrogen, oxygen or sulphur). There are also some low frequencies associated with deformations or oscillations of the molecular skeleton. This problem is discussed later, in the section dealing with characteristic group frequencies of the particular types of compounds. As an introduction, only those regions in which the characteristic bands of the most important groups occur will be considered here.

Bands appearing in the 3600–2700 cm^{-1} region arise from stretching vibrations of the —O—H, >N—H and ⋺CH groups [ν_{O-H} (IR), ν_{N-H} (IR, R), ν_{C-H} (R)*].

The 2600–2000 cm^{-1} region contain bands characteristic of the —S—H, ⋺C—D, >B—H, >P—H and ⋺Si—H stretching vibrations. In the similar but slightly narrower region (2300–2000 cm^{-1}) there also appear bands arising from the stretching vibrations of triple bonds —C≡C—[$\nu_{C\equiv C}$(R)], —C≡N [$\nu_{C\equiv N}$ (R, IR)] and —N≡N [$\nu_{N\equiv N}$ (IR)], and the antisymmetric vibrations of cumulene groups >C=C=C<, >C=C=O, —N=N=N, —N=C=O and —N=C=S, with a linear configuration of similar oscillators (cf. 4.1.3.1). The antisymmetric band is very strong in the infrared spectrum but weak in the Raman spectrum.

The 1900–1600 cm^{-1} region contains bands due to double bond stretching vibrations >C=O, >C=C<, >C=N— and —N=N— [$\nu_{C=O}$(IR, R), $\nu_{C=C}$(R), $\nu_{C=N}$(R, IR), $\nu_{N=N}$(R)], and the N—H deformation vibrations [δ_{N-H}(IR)].

---

* The letters R and IR denote bands that are strong in the Raman or infrared spectrum, respectively.

The 1600–1450 cm^{-1} region is characteristic of some bands associated with the stretching vibrations of aromatic rings (one-and-a-half bonds). The intensities of these bands vary according to the ring type and its substituents. Generally, the bands that are strong in the Raman spectrum are weak in the infrared spectrum, and vice versa. In the same region there are also bands corresponding to the antisymmetric stretching vibrations of the nitro (—NO_2) and carboxylate (—CO_2^-) groups [$\nu_{N—O}$ as and $\nu_{C—O}$ as (IR)]. The appropriate bands corresponding to the symmetric vibrations of these groups (strong in Raman) appear in the 1440–1330 cm^{-1} region.

In the 1450–1350 cm^{-1} region the deformation vibrations of the CH_3 and CH_2 groups (δ_{CH_3}, δ_{CH_2}) occur. These bands are usually weak in the Raman spectrum and of medium intensity in the infrared spectrum.

The 1350–650 cm^{-1} region is hard to decipher, since there are numerous bands due to skeletal and deformation vibrations which are strongly coupled. Some of these bands, in either the Raman or the infrared spectrum, can be identified by their intensity. Among the stronger Raman bands, some are due to symmetric breathing vibrations of rings (their band position depends on size and type of ring and its substitution), while others are associated with the $\nu_{S=O}$ and $\nu_{P=O}$ stretching vibrations, and the $\nu_{C—Cl}$ stretching vibrations (which are also strong in the infrared spectrum). In this region infrared spectra also show bands characteristic of out-of-plane rocking vibrations of hydrogen atoms attached to double bonds and aromatic rings ($\gamma_{C—H}$) as well as bands due to the stretching vibrations of pairs of heavier atoms, $\nu_{C—F}$, $\nu_{C—O}$, $\nu_{S—O}$, $\nu_{Si—O}$ and $\nu_{P—O}$.

The region below 650 cm^{-1} contains strong Raman bands due to stretching vibrations of bonds linking a carbon atom to a heavy atom (bromine, iodine, sulphur, silicon or a metal), or bonds linking two heavy atoms, e.g. S—S or Si—Si. There are also some bands associated with deformation vibrations, torsional vibrations or crystal lattice vibrations, which are useful only for detailed vibrational analysis.

4.1.3 Factors that influence group frequencies

Before discussing the characteristic frequencies of the particular functional groups of organic compounds, it is useful to list briefly some factors that tend to modify these frequencies. These factors are the spatial arrangement of groups in the molecule, the inductive and mesomeric effects of neighbouring groups, Fermi resonance and the physical state of the sample being examined.

4.1.3.1 Influence of the interatomic distances and spatial arrangement of groups

Two directly-coupled identical oscillators. If a diatomic fragment AB having a characteristic valence vibration is coupled with an identical fragment AB by a mutual atom B, a triatomic fragment ABA is formed. Such a triatomic fragment, which often occurs in organic molecules (e.g. methylene, amino, nitro, carboxylate, sulphonyl and dichloromethylene groups) is always characterized by two bands, one of which is due to the symmetric vibration and the other to the antisymmetric vibration. The difference between the frequency of the symmetric and the antisymmetric vibration is greatest when the angle ABA is 180° (Fig. 4.1).

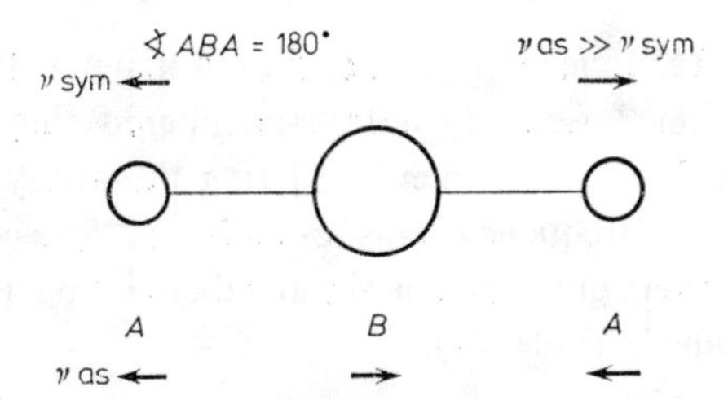

Fig. 4.1—The influence of a linear coupling of two identical oscillators on their fundamental frequencies.

The allene molecule can be used here as an example. The cumulene group >C=C=C< gives rise to an antisymmetric band at 1965 cm^{-1} in the infrared spectrum and a symmetric band at 1070 cm^{-1} in the Raman spectrum. The frequency of the isolated double bond stretching vibration C=C, however, is near 1640 cm^{-1}.

When the angle ABA changes from 180° to 90°, frequencies of the antisymmetric (νas) and the symmetric (νsym) vibrations approach each other in value, but always νas > νsym. If the angle ABA is smaller than 90°, the symmetric becomes higher than the antisymmetric frequency (Fig. 4.2). The bond angle also affects the characteristic frequencies of saturated rings. Although the number of valence vibrations in a ring is equal to the number of ring bonds, there are only two distinct bands. One of these is fully symmetric and due to the breathing vibration; it is very strong in the Raman spectrum (cf. Fig. 8.2, p. 184). The other is antisymmetric and very strong in the infrared spectrum, especially if the ring contains a heteroatom. A three-membered ring has a symmetric vibration

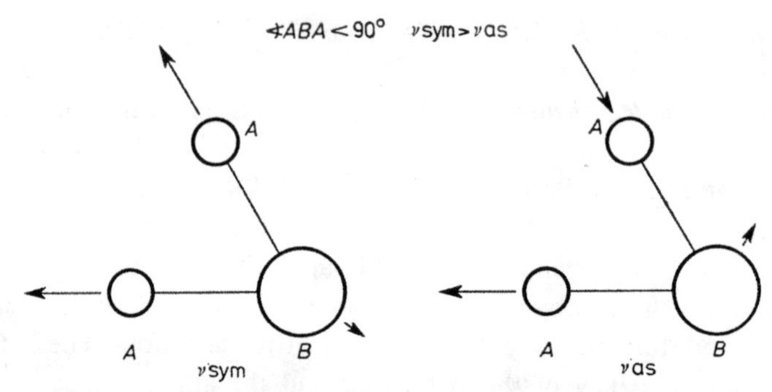

Fig. 4.2—The influence of the coupling of two identical oscillators positioned at an angle of less than 90°, on their fundamental frequencies.

with a band near 1250 cm^{-1} (νsym), and an antisymmetric vibration with a band near 820 cm^{-1} (νas). In a four-membered ring both frequencies are near 1000 cm^{-1}. In a five-membered ring the antisymmetric is higher than the symmetric frequency (νas $\approx$ 1060 cm^{-1}, νsym $\approx$ 900 cm^{-1}). The difference is even greater in a six-membered ring (νas $\approx$ 1120 cm^{-1} and νsym $\approx$ 820 cm^{-1}, Table 4.1).

Three directly-coupled identical oscillators. An A_3B group, for example the methyl or trichloromethyl group, has three valence vibrations (Fig. 4.3). Two of them are antisymmetric, one is symmetric. However, both antisymmetric vibrations have equal energy, so this vibration is doubly degenerate. Since the angles between the bonds of such a group are usually near 109°, the spectrum consists of two bands, an antisymmetric one of higher frequency and a symmetric one of lower frequency. In the valence bond region, this spectrum is similar in shape to that of an A_2B group, from which it differs only in the number of bands associated with bending vibrations (δ). The A_2B group has only one bending vibration, the so-called *scissoring* vibration, whereas the A_3B group has two bending vibrations, an antisymmetric (δas) of higher frequency, which is doubly degenerate, and a symmetric (δsym) of lower frequency.

Two indirectly-coupled identical oscillators. The ABBA type of oscillator coupling gives rise to an antisymmetric and a symmetric band, similar to those exhibited by the ABA group. The difference between the two frequencies is usually small; it depends on the difference between the masses of A and B, on the difference between the force constant of the AB bond and that of the BB bond, and on the spatial arrangement of the ABBA group. If A is a hydrogen atom and B an atom of a heavier element,

Table 4.1—Characteristic bands of saturated ring compounds.
(The symmetric stretching vibration (breathing) gives rise to an intense Raman band, the antisymmetric vibration is generally strong in the infrared spectrum)

Ring system	Band position, cm^{-1}	
	ν sym(R)	ν as(IR)
Cyclopropane	1185	
Aziridine	1210	
Oxirane[a]	1270	840
Thiirane[b]	1120	
Cyclobutane	1000	920
Oxetane[c]	1028	980
Cyclopentane	890	
Tetrahydrofuran	915	1070
Tetrahydrothiophene	690	
Pyrrolidine	900	1078
Cyclohexane	800	
Tetrahydropyran	815	1098
Piperidine	817	
1,4-Dioxane	835	1120
1,3,5-Trioxane	960	1175
Cycloheptane	735	
Cyclo-octane	705	

[a] Ethylene oxide, [b] ethylene sulphide, [c] trimethylene oxide.

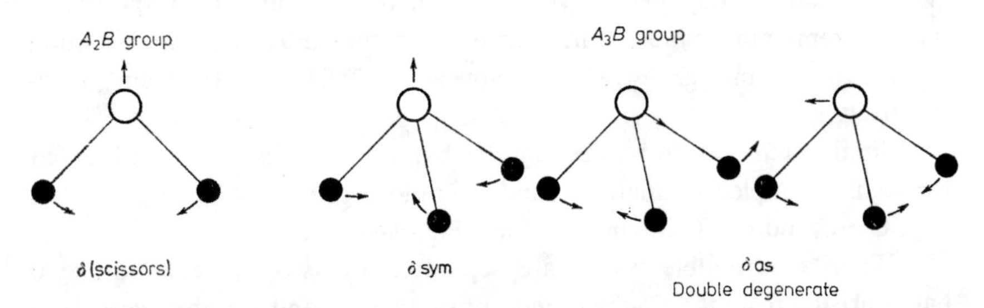

Fig. 4.3—Comparison between the deformation vibrations of the A_2B group and the A_3B group.

then the difference between the two frequencies is small, and usually inconspicuous in the spectrum. If A and B are linked by a double bond and BB by a single bond, the two bands are distinctly separated in the spectrum. In this case the spatial arrangement of both oscillators in the molecule can be determined by comparing the intensities of the bands in the Raman and infrared spectra. If the AB bond has sp^2 hybridization then because of π-electron coupling, the double bonds lie in one plane and the group can exist in the *Z* conformation (I) or the *E* conformation (II). Form II gives rise to a higher antisymmetric frequency and a lower symmetric frequency positions of the antisymmetric and symmetric vibrations are reversed — it is the case of form I (Fig. 4.4). Butadiene, for example,

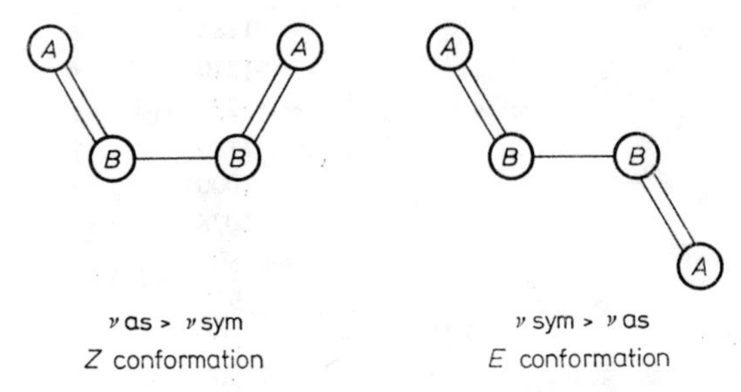

Fig. 4.4—The influence of spatial arrangements of two conjugated double bonds on their valence vibration frequencies.

which exists only in the *E* conformation, has a strong symmetric band at 1463 cm^{-1} in the Raman spectrum and a strong antisymmetric band at 1600 cm^{-1} in the infrared spectrum. On the other hand, 2,4-dimethylpentadiene molecules exist in both conformations and give rise to four bands in the Raman spectrum. Two at 1642 and 1629 cm^{-1} are strong and are related to symmetric vibrations of forms I and II, respectively; the two remaining bands, which arise from the antisymmetric vibration of the two forms, are weak and appear at 1604 and 1659 cm^{-1}, respectively.

Similar relationships affecting the band intensities can be observed for similar coupled oscillators situated in close proximity to one another (e.g. C=C and C=O in vinyl ketones, Fig. 9.44, p. 239).

When two double bonds are separated by two single bonds, two bands arising from the double-bond antisymmetric and symmetric stretchings also appear in the spectrum. Spectra of carboxylic acid anhydrides and β-diketones may be given as examples (see Section 4.2.3.3).

Coupled oscillators with different force constants. Although the C=C and C=O stretching vibrations are characteristic, it should be remembered that they are not completely localized. Their frequencies may vary with alteration of the angles between the groups and other atoms, though the double-bond force constant does not change markedly (cf. for example the frequencies of the endocyclic and exocyclic bands $\nu_{C=C}$ in Table 4.6,

p. 100, and those of cyclic ketones in Table 4.22, p. 121). The effect is even stronger in non-substituted or monosubstituted amides (see Section 4.2.3.3).

4.1.3.2 Influence of Fermi resonance

If the frequency of an overtone ($2\nu_1$) or a combination tone ($\nu_1+\nu_2$) is almost the same as a fundamental frequency (e.g. the characteristic vibration) and if the vibrations are of the same symmetry type, then the phenomenon of Fermi resonance may occur (cf. Section 8.5, p. 185). It gives rise to two new bands, one at a higher frequency and the other at a lower frequency than the primary transition frequencies. Simultaneously, another interesting phenomenon is observed in which the intensity of the combination tone (or overtone) band, which is usually weak, increases significantly at the cost of the fundamental band intensity. In the extreme case both bands have the same intensity. These two factors, i.e. the band position and intensity, which differ from those normally observed, may thus lead to an erroneous interpretation of the spectrum. For example, in the Raman spectrum of carbon dioxide there are two strong bands at 1388 and 1285 cm^{-1}, instead of one stretching band at 1336 cm^{-1}. This is caused by Fermi resonance between the tone that is associated with the excitation of the characteristic group frequency ($\nu_{C=O}$sym) and the overtone due to the deformation frequency (δ), which appears at 667 cm^{-1}.

4.1.3.3 Influence of the inductive, mesomeric and field effect of neighbouring groups

Groups situated close to an AB oscillator having a characteristic frequency can influence the electron density between atoms A and B and so change the AB bond force constant, by the inductive or mesomeric neighbouring group effect. The most characteristic frequency shift occurs for multiply-bonded atoms.

The influence of the inductive effect is exemplified by the SO group, which may be represented by two resonance structures:

$$R_2S{=}O \leftrightarrow R_2\overset{\oplus}{S}{-}\overset{\ominus}{O}$$

(1) (2)

The sulphur–oxygen stretching frequency in dimethylsulphoxide is 1055 cm^{-1}. If the carbon atoms bonded to the sulphur atom are replaced by strongly electronegative atoms, e.g. oxygen or chlorine, electrons move from the oxygen to the sulphur atom, resulting in increased importance of structure (1). The sulphur–oxygen bond acquires more double-bond character, which results in an increase in the force constant. For instance, the $\nu_{S—O}$ band in the spectrum of diethyl sulphite appears at 1210 cm^{-1}. In carbonyl compounds the double-bond character of the carbon–oxygen bond is influenced not only by the inductive effect of the neighbouring group but also by the mesomeric effect, which alters the C=O force constant. A grouping in which a substituent X is bonded to a carbonyl group may be represented by the three following resonance structures:

$$\underset{(1)}{R—\overset{\overset{O}{\|}}{C}—X} \qquad \underset{(2)}{R—\underset{\oplus}{\overset{\overset{O^{\ominus}}{|}}{C}}—X} \qquad \underset{(3)}{R—\overset{\overset{O^{\ominus}}{|}}{C}=X^{\oplus}}$$

If X is a carbon atom in, for example, acetone, the $\nu_{C=O}$ valence frequency is 1715 cm^{-1}. If the carbon atom is replaced by a more electronegative chlorine atom, the frequency increases to 1805 cm^{-1} because of the increasing contribution of structure (1). However, in esters (e.g. ethyl acetate) the $\nu_{C=O}$ frequency decreases to 1750 cm^{-1}, because the mesomeric effect also has an influence which opposes that due to the electronegative character of oxygen. Hence, the greater contribution of resonance structure (3) results in decrease of the carbon–oxygen bond force constant to below the value for the acid chloride. The greatest contribution of structure (3) to the resonance hybrid may be expected when X is a nitrogen atom. Nitrogen, being less electronegative than oxygen, diminishes the contribution of structure (1) to the resonance hybrid, so that the carbon–oxygen force constant is very low. Accordingly, the $\nu_{C=O}$ frequency in *N,N*-dimethylacetamide is only 1680 cm^{-1}.

In the spectra of halocarbonyl compounds two bands are usually observed. The distance between the band of higher frequency and the normal position of the carbonyl frequency is about 20 cm^{-1}. These bands correspond to the two conformations of the molecule. The conformation in which the halogen atom is synclinal to the oxygen atom gives rise to the lower frequency band; the conformation in which the halogen atom is synperiplanar to the oxygen atom gives rise to the band of higher frequency. The spatial electrostatic interaction of the two neighbouring negatively charged sites $O^{\delta-}=C—C—X^{\delta-}$ results in mutual repulsion of the negative charges, thus increasing the carbon–oxygen bond order. The contribution of structure (1) increases and the force constant of the C—X bond alters. The result of this interaction is called the *field effect*. Chloroacetone is used here as an example. The C═O stretching vibrations of the two conformations give rise to bands at 1725 and 1740 cm^{-1}, and the C—Cl stretching vibrations to bands at 765 and 730 cm^{-1}.

4.1.3.4 Influence of physical state, polarity of the environment and the formation of hydrogen bonds

Undisturbed normal modes can be observed for molecules in the gaseous state. A change to the condensed phase with intermolecular interactions (van der Waals forces), formation of hydrogen bonds, and production of complexes, leads to a lowering of the characteristic frequencies, compared with those observed in the gaseous state. For gaseous acetone, for example, the $\nu_{C=O}$ frequency is 1740 cm^{-1} whereas for liquid acetone it is 1715 cm^{-1}. The greatest decrease in frequency may be observed for valence vibrations of strongly polar groups, for example the O—H and C═O groups. Therefore non-polar solvents should be used, if possible, in spectral analysis.

Strong hydrogen bonds can be formed in compounds that contain the O—H and N—H groups. Atoms with significant electron affinity and lone pairs of electrons behave as proton acceptors. If a hydrogen bond is formed, the X—H bond suffers a loss in strength, so the force constant decreases. In the spectrum of 4-chlorophenol the band corresponding to the associated hydroxyl group appears at 3250 cm^{-1}, whereas the band corresponding to the free hydroxyl group in a dilute solution in carbon tetrachloride appears at 3610 cm^{-1}. If the oxygen atom of a carbonyl group participates in a hydrogen bond, its valence frequency will decrease. For example, the vibration frequency $\nu_{C=O}$, of methyl benzoate gives rise to a band at 1730 cm^{-1}, whereas the corresponding band in the methyl salicylate spectrum appears at 1680 cm^{-1}, because methyl salicylate contains a strong intramolecular hydrogen bond. The formation of a hydrogen bond by the X—H group hinders its deformation vibrations and thus increases their frequency. The δ_{NH_2} vibration in pure hydroxylamine gives rise to a band at 1635 cm^{-1}, but this band appears at 1600 cm^{-1} in the spectrum of its dilute solution in carbon tetrachloride.

Transitions from the liquid to the solid state may affect the spectrum in two ways. Often a simpler spectrum is observed (some bands disappear) because of the smaller number of conformations in which the molecules can exist. On the other hand, local electric fields in the crystal lattice can cause splitting of some bands (so-called *crystal field splitting*) and can also affect the band intensity.

4.2 CHARACTERISTIC FREQUENCIES

To enable the reader to visualize the relations between the appearance of a vibrational spectrum and the structure of the compound to be identified, a spectral atlas is given at the end of this book (Chapter 9). It is composed of spectra which include the characteristic bands associated with groups discussed in this chapter. The band positions are given in descriptions of the spectra. In the following sections, which deal with particular groups and their characteristic band regions, references are made to those figures containing bands of the groups in question (Figs. 9.1–9.66). Since the same structural element may appear in several compounds, the figure numbers for the appropriate spectra are given as footnotes in the tables of characteristic frequencies.

4.2.1 Hydrocarbons

4.2.1.1 Aliphatic groups in alkanes

Aliphatic groups, which are the only constituents of saturated hydrocarbons, are characterized by vibrations of the terminal methyl and chain methylene groups. In a methyl group there are six types of vibrations: stretching antisymmetric ($\nu_{C—H}$ as); stretching symmetric ($\nu_{C—H}$ sym); deformation antisymmetric ($\delta_{C—H}$ as); deformation symmetric ($\delta_{C—H}$ sym); rocking (ϱ) and twisting (τ). The methylene group also has six characteristic vibrations: stretching antisymmetric ($\nu_{C—H}$ as); stretching symmetric ($\nu_{C—H}$ sym); deformation, also spoken of as scissoring, (δ); wagging (ω); twisting (τ) and rocking (ϱ) (cf. Fig. 8.2, p. 184). For identification, the stretching and deformation bands are mostly used, the other band types being used only occasionally. Bands arising from the vibrations of carbon–carbon bonds ($\nu_{C—C}$) are not very characteristic and are therefore useful only as an additional piece of information for the identification of the carbon chain.

The bands characteristic of aliphatic groups are given in Table 4.2. Bands due to the C—H stretch are strong in the Raman spectrum, the more intense band being associated with the symmetric vibration. These bands are weak in the infrared spectrum and the ratio of their intensities is reversed, i.e. $I_{\nu as} > I_{\nu sym}$. The characteristic regions usually contain additional bands arising from various forms of the C—H oscillator couplings and from Fermi resonance. A band near 2960 cm^{-1} allows the methyl group to be distinguished from the methylene group. Deformation vibra-

Table 4.2—Bands characteristic of aliphatic groups

Group	Type of vibration	Band position, cm^{-1}	Intensity* R	Intensity* IR
CH_3	ν as	2960 ± 10	s	m
	ν sym	2870 ± 10	s	m
	δ as	1460	w	m
	δ sym	1375	vw	m
	ϱ	1135		
CH	ν	2890 ± 10	w	vw
	δ	1340	w	w
$(CH_2)_n$[a]	ω	1300	s	—
	ϱ	720	—	m
	ν_{C-C} {2 bands	1100–1000		
		(1075)†		
	{3 or 4 bands	900–800		
	torsion	420–150		
R–C(R)–C–[b] (—C—C— with R above and R below)	$\nu_{skeletal}$	1170–1160	m	—
		970–960	m	—
		810	m	—
CH_2	ν as	2925 ± 10	s	m
	ν sym	2850 ± 10	s	m
	δ	1465	m	m
	ω	1300	w	
$(CH_3)_2CH$[c]	δ	1380	—	m
	δ	1370	—	m
	$\nu_{skeletal}$	1170	m	m
		1155	m	m
$(CH_3)_3C$[d]	δ	1395	—	w
	δ	1370	—	m
	$\nu_{skeletal}$	1250		m
		1205		m
		750–650	s	w
		930	m	w

* vs—very strong, s—strong, m—medium, w—weak, vw—very weak, va—variable, br—broad bands.
† One usually at 1075.
[a] Cf. bands in Figs. 9.2, 9.17, 9.27, 9.29, 9.47, 9.51.
[b] Cf. band in Fig. 9.2.
[c] Cf. band in Fig. 9.1.
[d] Cf. bands in Figs. 9.2, 9.5, 9.50.

tions give rise to weak bands in the Raman spectrum (the δ_{sym} methylene band is very weak) and medium strength bands in the infrared spectrum. Isopropyl and tert-butyl groups exhibit characteristic doublets of the δsym bands, accompanied by skeletal bands also located in characteristic positions of the spectrum. If there are several neighbouring CH_2 groups in a molecule, the characteristic band arising from the wagging vibration (ω_{CH_2}) appears at 1300 cm^{-1} in the Raman spectrum. Longer hydrocarbon chains are additionally characterized by a series of bands (3 or 4) due to vibrations of carbon–carbon bonds ($\nu_{C—C}$) in the ranges 900–800 cm^{-1} and 1100–1000 cm^{-1} (normally a doublet near 1075 cm^{-1}) in the Raman spectrum. A band due to the methylene rocking vibration (ϱ_{CH_2}) near 720 cm^{-1} is seen in their infrared spectrum. Moreover, there are some strong bands in the Raman spectrum, in the 450–150 cm^{-1} region, due to carbon-chain deformation vibrations. Their frequencies increase with increasing length of the chain. The spectrum of decane may serve as a typical example (Fig. 9.1).

Branched-chain hydrocarbons usually have more bands in their spectra than do straight-chain hydrocarbons. Characteristic bands of branched chains occur in the Raman spectrum in the regions 1170–1160 cm^{-1} and 970–960 cm^{-1} and also near 810 cm^{-1} (cf. spectrum in Fig. 9.2).

Spectra of straight-chain hydrocarbons have no strong bands between 900 and 800 cm^{-1}, but the spectra of branched chains exhibit neither series of bands between 900 and 800 cm^{-1}, nor the band at 1300 cm^{-1} which is a good indication of the presence of a large number of CH_2 groups.

4.2.1.2 Cycloalkanes

The C—H stretching frequency of cycloalkanes increases with decreasing ring size from six-membered to three-membered, whereas the deformation frequency decreases. This can be seen by comparing the spectra given in Figs. 9.3, 9.16, 9.24, 9.31 (Table 4.3). Characteristic bands in the Raman spectrum, arising from the ring breathing vibration, can indicate the presence of a saturated ring and also its size (see Table 4.1, p. 87). The frequency of the breathing vibration varies with ring substitution (cf. Table 4.3 and the spectra mentioned).

4.2.1.3 Aliphatic groups in various systems

The characteristic frequencies of the methyl and methylene groups are slightly but significantly shifted if these groups are linked to a carbon atom of an unsaturated group or to a hetero atom, rather than to an aliphatic carbon atom (Table 4.4). These shifts are useful for identification purposes.

Table 4.3—Band positions in the spectra of cycloalkanes

Ring type	Type of vibration	Band position, cm^{-1}
C_3H_6 [a]	$\nu_{C—H}$ as	3100–3090
	$\nu_{C—H}$ sym	3040–3020
	δ_{CH_2}	1445
	$\nu_{C—C}$ sym ring vibration	1300–1200
C_3H_8 [b]	$\nu_{C—H}$ as	2990–2975
	$\nu_{C—H}$ sym	2900–2885
	δ_{CH_2}	1445
	$\nu_{C—C}$ sym ring vibration	940–860
C_5H_{10} [c]	$\nu_{C—H}$ as	2960–2950
	$\nu_{C—H}$ sym	2865–2850
	δ_{CH_2}	1460–1450
	$\nu_{C—C}$ sym ring vibration	900–820
C_6H_{12} [d]	$\nu_{C—H}$ as	2930–2915
	$\nu_{C—H}$ sym	2885
	δ_{CH_2}	1445
	$\nu_{C—C}$ sym ring vibration	780–740 [e]

[a] Cf. spectrum in Fig. 9.3.
[b] Cf. spectrum in Fig. 9.31.
[c] Cf. spectrum in Fig. 9.24.
[d] Cf. spectra in Figs. 9.16, 9.38.
[e] Exceptionally, even down to 705 cm^{-1} for the 1.1-dimethyl derivative.

Halogens, the carbonyl group and the nitro group cause the valence frequency of neighbouring C—H groups to increase. On the other hand, the nitrogen atoms of amines shift the valence bands to lower frequencies, so the band due to the symmetric carbon–hydrogen stretch ($\nu_{C—H}$) appears in a region where it is no longer obscured by other $\nu_{C—H}$ bands. Every unsaturated group and every hetero atom causes the deformation frequencies of neighbouring CH_3 and CH_2 groups to decrease. An exception is oxygen, which shifts the methyl group symmetric vibration to higher frequencies. The decrease in frequency of deformation vibrations depends markedly on the mass of the neighbouring hetero atom.

4.2.1.4 Alkenes

The carbon–carbon double bond is characterized by a stretching vibration, which gives rise to a strong or very strong band near 1650 cm^{-1} in the Raman spectrum. Its intensity in the infrared spectrum is low and decreases further with increasing molecular symmetry. If, however, the double bond is linked to

Table 4.4— Band positions for aliphatic groups joined to groups containing multiple bonds or hetero atoms

Group	Type of vibration	Band position, cm^{-1}	Group	Type of vibration	Band position, cm^{-1}
CH_3—Ar	ν as	2925	CH_2—Ar	δ	1440±10
	ν sym	2865±10	CH_2—C=O	ν	3000–2900
	δ as	1425±15	CH_2—NO_2	ν	3000
	δ sym	1360		δ	1440
CH_3—C=O	ν as	3000–2900	CH_2—N	ν as	2930
	ν sym			δ	1460±15
	δ as	1425		ν sym	2790±30
	δ sym	1360	CH_2—O	δ	1460±15
CH_3—NR	ν sym	2790±10	CH_2—Cl	ν	3000–2950
CH_3—NAr	ν sym	2810±10		δ	1445±10
	δ	1425±16		ω	1270±30
CH_3—O	ν as	2980±20	CH_2—Br	ω	1230
	ν sym	2880±10	CH_2—I	ω	1170
	δ as, sym	1450±10	CH_2—S	δ	1430±10
CH_3—S	δ as	1425±15		ω	1250±20
	δ sym	1310±20	CH_2–Si	δ	1410
CH_3—Si	δ as	1420±15	CH_2—P	δ	1425±15
	δ sym	1260±10			
CH_3—P	δ sym	1300±20			

electronegative substituents, or those possessing multiple bonds, the band intensity increases. For endocyclic and exocyclic bonds the carbon–carbon stretching frequency ($\nu_{C=C}$) varies specifically, depending on the ring size (Table 4.5). In tetra-, tri- and *trans*-dialkyl-substituted alkenes the stretching frequency of the carbon–carbon double bond ($\nu_{C=C}$) is slightly higher than that in alkenes having two geminal substituents in the *cis*-form, or in alkenes with only one substituent (vinyl group). If the C=C bond is conjugated with another multiple bond (C=C, C=O, C=N, C≡C, C≡N) or with an aryl group, then the stretching frequency ($\nu_{C=C}$) is shifted 30–50 cm^{-1} to lower frequencies because of the diminished carbon–carbon bond order (Figs. 9.4 and 9.44). The spatial arrangement of the conjugated bond system (Fig. 4.4) can also influence the frequency. A decrease in the vibration frequency ($\nu_{C=C}$) is observed if a halogen atom (except fluorine, which shifts the frequency to higher values) or another heavy atom (e.g. sulphur or phosphorus) is adjacent to the double bond. For instance the C=C stretching frequency in *Z*-1,2-dichloroethylene is 1590 cm^{-1}, whereas for 1,1-difluoroethylene it increases to 1730 cm^{-1}. The number of bands in the polyene spectrum can be equal to the number of double bonds in the compound. Not all the bands are visible, however, in both the Raman and infrared spectra because the symmetric vibration tends to give rise to strong bands in the Raman spectrum, whereas the antisymmetric vibration gives rise to strong bands in the infrared spectrum. In spectra of conjugated dienes there are two bands, one near 1650 cm^{-1} and the other near 1600 cm^{-1}. The intensity of these bands may vary depending on the spatial arrangement. The mechanical coupling is strongest for a linear arrangement of the double bonds. Thus the antisymmetric vibration in cumulenes gives rise to a band near 1970 cm^{-1} in the infrared spectrum and the symmetric vibration to a band near 1070 cm^{-1} in the Raman spectrum (cf. the triene spectrum, Fig. 9.5).

If there are hydrogen atoms bonded to the olefinic carbon atom, more information about the configuration of atoms around this bond can be obtained. Bands arising from the carbon–hydrogen bond stretching vibrations ($\nu_{=C-H}$) appear in the region above 3000 cm^{-1} (up to 3100 cm^{-1}), so they can be distinguished from the corresponding bands of alkanes. This region also contains, however, bands due to aromatic hydrocarbons, some halogen compounds and three-membered rings. Bands due to the C—H in-plane bending vibration (δ) are of little value for the purpose of identification. Much more characteristic are the bands due to the C—H out-of-plane vibrations (γ) of double bonds. The position of these bands,

Table 4.5—Band positions $\nu_{C=C}$ for unsaturated groups

Group	Band position, cm^{-1}
Ar—C=C[a]	1625
(—C=C—)$_2$	1650, 1600
C=C—C≡C	1615
C=C=C	1980 as
	1070 sym
C=C=C=C[b]	2080 sym
	880 sym
F—C=C—F	1710–1695
C=CF_2	1740–1715
F—C=CF_2	1795
F_2C=CF_2	1870
C=C—I	1580
Cyclopropene	1640
Cyclobutene	1565
1,2-Dialkylcyclobutene	1685
Cyclopentene	1610
Cyclohexene	1645
1,2-Dialkylcyclohexene	1680
C=C—Cl[c]	1605
C=CCl_2	1615
Cl—C=C—Cl	1590–1575
Cl—C=CCl_2	1590
Cl_2C=CCl_2	1570
I—C=C—I	1540
(C_3H_4)=CH_2	1780
(C_4H_6)=CH_2	1680
(C_5H_8)=CH_2	1660
(C_6H_{10})=CH_2	1650
(C_5H_8)=CR_2	1690
(C_6H_{10})=CR_2	1670
C=C—Br	1595
C=CBr_2	1595
Br—C=C—Br	1580
Br—C=CBr_2	1550
Br_2C=CBr_2	1545
I_2C=CI_2	1465

[a] Cf. spectrum in Fig. 9.4.
[b] Cf. spectrum in Fig. 9.5.
[c] Cf. spectrum in Fig. 9.44.

which are in practice visible only in the infrared spectrum, provides a good method for the identification of the configuration of groups (*cis* or *trans*) attached to a double bond (cf. Table 4.6 and Figs. 9.4, 9.44 and 9.53).

4.2.1.5 Alkynes

The C≡C stretching mode gives rise to a very strong band in the Raman spectrum, whereas in the infrared spectrum it is weak or even invisible, especially if the triple bond is situated within a chain (cf. the spectrum of diphenylacetylene, Fig. 9.7). In spectra of disubstituted acetylenes two bands normally appear, near 2230 and 2300 cm^{-1}, due to Fermi resonance between the $\nu_{C\equiv C}$ vibration and the ν_{C-C} overtone (Fig. 9.7). In monosubstituted acetylene derivatives, the carbon–hydrogen stretching vibration (ν_{C-H}) gives a band which appears near 3300 cm^{-1}. It is quite intense in the infrared spectrum but of medium intensity in the Raman spectrum (Fig. 9.6). This band is sharper than other bands which appear in the same region (ν_{O-H}, ν_{N-H}) and are strong in the infrared spectrum. Moreover, these bands are shifted if the sample is diluted with a non-polar solvent. The ≡C—H group also gives rise to a band in the 650–600 cm^{-1} region, associated with the ω_{C-H} vibration (Fig. 9.6, Table 4.7). This band is strong in the infrared spectrum and of medium strength in the Raman spectrum.

4.2.1.6 Aromatic hydrocarbons

Benzene and its derivatives. Bands occurring in the spectra of aromatic hydrocarbons allow confirmation of the presence of the aromatic ring and the type of substitution. These bands are ascribed to the C—H stretching and bending modes, ring stretching and bending modes and composite tones. Bands arising from the ν_{C-H} vibrations appear in the range 3100–3000 cm^{-1}, which is characteristic of unsaturated groupings. There are generally three bands (near 3060, 3030 and 3010 cm^{-1}), their number decreasing with increasing number of substituents on the ring. The bands are of medium intensity (one of them is normally stronger than the others) in the Raman spectrum but weak in the infrared spectrum. In spectra of benzene derivatives four carbon–carbon stretching bands can occur in the 1630–1450 cm^{-1} region, near 1600, 1580, 1500 and 1450 cm^{-1}. The band intensities depend on the positions of the substituents and on their properties as electron donors or electron acceptors. If the molecule contains aliphatic groups, the fourth band is generally difficult to identify because there are bands due to this group (δ_{CH_2} and δ_{CH_3}) in the same region. The first band, which is generally but not always accompanied by the

Table 4.6—Bands characteristic of the olefinic group in various molecular environments

Molecule	$\nu_{C=C}$ vibration			ν_{C-H} vibration			δ_{C-H} vibration			γ_{C-H} vibration		
	Band position, cm^{-1}	Intensity R	Intensity IR	Band position, cm^{-1}	Intensity R	Intensity IR	Band position, cm^{-1}	Intensity R	Intensity IR	Band position, cm^{-1}	Intensity[a] R	Intensity[a] IR
R—CH=CH_2	1645±5	s	m	3085±10 as	m	w	1410±10	m	s	990		
				3030±10 sym	m	w	1300±5	m	w	910		
R_2C=CH_2 (R\\C=CH_2 /R)	1655±10	s	m	3085	m	w	1410±10	m	s	890		
cis-RCH=CHR[b]	1660	s	w	3020	m	w	1410±10		m	700±30		
trans-RCH=CHR[c]	1670	vs	vw	3020	m	w	1300±10		m	975±15		
R_2C=CHR	1670	vs	w	3020	w	vw	1345		w	820±20		
R_2C=CR_2	1670	vs	—									

[a] Intense only in the infrared. The γ_{C-H} frequency of the =CH_2 fragment of the vinyl group is influenced by the mesomeric effect of neighbouring groups. This frequency is decreased if electrons are donated to the vinyl group, and increased if electrons are accepted from the vinyl group. The γ_{C-H} frequency of the hydrogen atoms in the *trans*-position H\C=C\H is influenced by the inductive effect of neighbouring groups, and decreases with increasing electronegativity of these groups.

[b] Cf. spectrum in Fig. 9.53.

[c] Cf. spectrum in Fig. 9.44.

Table 4.7—Bands characteristic of the acetylenic group

Group	$\nu_{C\equiv C}$ vibration			ν_{C-H} vibration			ω_{C-H} vibration		
	Band position, cm^{-1}	Intensity R	Intensity IR	Band position, cm^{-1}	Intensity R	Intensity IR	Band position, cm^{-1}	Intensity R	Intensity IR
C≡C—H[a]	2120±20	vs	w	3300±30	m	s	650–600	w	s
R—C≡C—R[b]	2225±30	vs	—						
	2300±10	s							

[a] Cf. spectrum in Fig. 9.6.
[b] Cf. spectrum in Fig. 9.7.

second one, is strong in the Raman spectrum and weak in the infrared spectrum. The third, near 1500 cm^{-1}, is normally not visible in the Raman spectrum, whereas it is intense in the infrared spectrum (except for the spectra of carbonyl compounds).

In the spectra of *para*-derivatives the first and third bands appear at 10–20 cm^{-1} higher frequencies and the fourth band falls at lower frequency than the corresponding bands in *ortho*- and *meta*-derivatives (Table 4.8).

Table 4.8—Positions of bands characteristic of benzene and its derivatives

Type of vibration	Band position, cm^{-1}	Intensity	
		R	IR
$\nu_{C—H}$	3080–3010	m	w
$\delta_{C—H}$	1270–980	va	va
$\gamma_{C—H}$	900–720	vw	vs
$\nu_{C=C}$ sym	1005–990	vs	vw
$\nu_{C=C}$ (1)	1620–1600	s	w
$\nu_{C=C}$ (2)	1590–1570	s	w
$\nu_{C=C}$ (3)	1510–1480	w	vs
$\nu_{C=C}$ (4)	1450–1430	m	s

In the Raman spectra of unsubstituted benzene and its mono- (Fig. 9.6), 1,3- (Fig. 9.9) and 1,3,5-derivatives, in the 1005–990 cm^{-1} region there is a band ascribed to the symmetric (breathing) ring vibration. This band is very strong. In the spectra of 1,4-derivatives a strong band appears in the 830–720 cm^{-1} region (see, for example, the spectra in Figs. 9.4 and 9.10). Some remaining skeletal vibrations are coupled with vibrations of the substituent and are therefore rarely used for identification (Table 4.9). In the infrared spectra of monosubstituted, 1,3,5- and 1,2,3-benzene derivatives, a strong band ascribed to the ring deformation vibration appears in the 710–690 cm^{-1} region. Bands due to the ring deformation vibrations, the frequency of which varies with the type of substitution, also appear in the regions 650–620, 580–430 and near 450 cm^{-1}. The C—H in-plane bending vibration (δ) gives rise to a series of sharp bands of varying strength in the 1300–960 cm^{-1} region, generally near 1270, 1160, 1100, 1070, 1030 and 980 cm^{-1} (Table 4.10). These bands confirm the presence of an aromatic ring. Of value in the identification of monosubstituted benzene derivatives is a band in the Raman spectrum, near 1030 cm^{-1} (Fig. 9.7), accompanying the very strong breathing band at 1000 cm^{-1}. Similarly, the band characteristic of the 1,2-benzene derivatives appears in the Raman

Table 4.9—Bands characteristic of the benzene ring substitution type*

Substitution type	Vibration	Band position, cm^{-1}
Unsubstituted	$\nu_{C=C}$ sym	995 R
	ν_{C-H}	670 IR
Monosubstituted[a]	δ_{C-H}	1030 R
	$\nu_{C=C}$ sym	1000 R
	ν_{C-H}	770–730 IR
	δ_{C-C}	710–690 IR
1,2-Disubstituted[b]	δ_{C-H}	1040 R
	γ_{C-H}	770–730 IR
	$\nu_{skeletal}$	1230–1210[c] R
		740–715[c] R
		680–650 R
		600–560 R
		560–540 R
1,3-Disubstituted[d]	$\nu_{C=C}$ sym	1000 R
	γ_{C-H}	810–750 IR
	δ_{C-C}	720–680 IR
	$\nu_{skeletal}$	1260–1210 R
		1180–1160 R
		740–700 R
		540–520 R
1,4-Disubstituted[e]	$\nu_{C=C}$	830–720 R
	γ_{C-H}	860–800 IR
	$\nu_{skeletal}$	1230–1200[c] R
		1180–1150[f] R
		650–630 R
1,2,3-Trisubstituted[g]	δ_{C-H}	1100–1050 R
	γ_{C-H}	810–750 IR
	δ_{C-C}	740–705 IR
	δ_{C-C}	670–500 R
1,2,4-Trisubstituted	γ_{C-H}	890–870 IR
	γ_{C-H}	820–805 IR
	δ_{C-C}	750–650 R
	$\nu_{skeletal}$	1280–1200 R
1,3,5-Trisubstituted	$\nu_{C=C}$ sym	1000 R
	γ_{C-H}	880–820 IR
	δ_{C-C}	730–675 IR
	δ_{C-C}	570–510 R

* R—bands intense in the Raman spectrum, IR—bands intense in the infrared spectrum.
[a] Cf. bands in Figs. 9.5–9.7, 9.20, 9.21. [b] Cf. bands in Figs. 9.8, 9.14, 9.25, 9.41.
[c] Characteristic of dialkyl substituted benzenes. [d] Cf. bands in Figs. 9.9, 9.26.
[e] Cf. bands in Figs. 9.4, 9.10, 9.19, 9.23, 9.33. [f] Two bands
[g] Cf. band in Fig. 9.43.

Table 4.10—Position of bands corresponding to the δ_{C-H} vibrations of benzene derivatives ($\pm 10\ cm^{-1}$)

Monosubstituted derivatives	Disubstituted *o*-derivatives	Disubstituted *m*-derivatives	Disubstituted *p*-derivatives
1240	1270	1280	1260
1180	1160	1160	1175
1160	—	—	—
1070	1125	1095	1120
1030[a]	1040[a]	1075	1010

[a] Strong Raman bands characteristic of the substitution type.

spectrum near 1040 cm^{-1} (Fig. 9.8). The position of the C—H out-of-plane bending (γ) bands (Table 4.11) depends on the number of neighbouring hydrogen atoms bonded to a ring, thus allowing identification of the type of ring substitution. These bands are very strong in the infrared spectrum but weak in the Raman spectrum. The spatial arrangement of substituents bonded to a benzene ring can also be recognized by means of weak bands occurring in the 2000–1700 cm^{-1} region in the infrared spectrum (Fig. 4.5). These bands are ascribed to composite tones and overtones and their patterns are characteristic of substitution type.

Aromatic hydrocarbons with condensed rings. The spectra of polycyclic hydrocarbons, like the spectra of benzene derivatives, have ν_{C-H} bands in the 3100–3000 cm^{-1} region, band series ascribed to the $\nu_{C=C}$ vibrations in the 1630–1350 cm^{-1} region, and γ_{C-H} bands in the 900–700 cm^{-1} region. It is possible to identify the substitution type from the number and position of γ_{C-H} bands in the infrared spectrum, since the band position depends (as for benzene derivatives) on the number of neighbouring hydrogen atoms. In the spectra of hydrocarbons with condensed rings many more bands generally appear than in the spectra of benzene derivatives. The

Table 4.11—Positions of strong infrared bands associated with the γ_{C-H} vibrations

Number of neighbouring hydrogen atoms	Band position, cm^{-1}
5, 4	770–720
3	810–750
2	860–800
1	900–860

strong band near 1000 cm^{-1}, characteristic of a number of benzene derivatives, is absent in the Raman spectra of derivatives of naphthalene, anthracene and phenanthrene. In the Raman spectra of naphthalene derivatives there are two strong bands arising from the skeletal vibration, one near 1575 cm^{-1} and the other in the 1390–1350 cm^{-1} region. In the infrared spectrum, strong bands appear near 1600 (often a doublet), near 1505 and in the region 1390–1350 cm^{-1}. Their positions differ in general from those in the Raman spectrum (cf. the spectrum of chloronaphthalene, Fig. 9.49). For example, in unsubstituted naphthalene there are fairly intense bands at 1575, 1460 and 1380 cm^{-1} in the Raman spectrum and at 1595, 1505 and 1390 cm^{-1} in the infrared spectrum. Anthracene derivatives (Fig. 9.30)

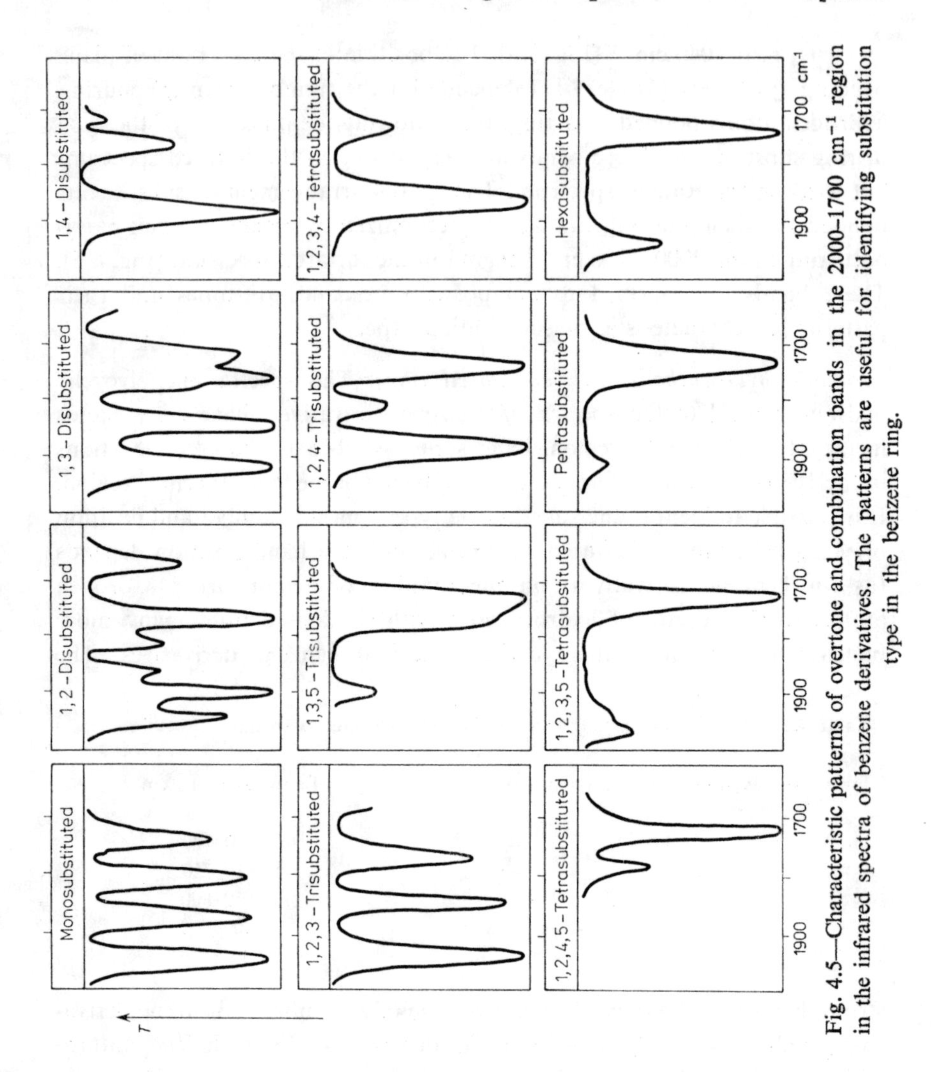

Fig. 4.5—Characteristic patterns of overtone and combination bands in the 2000–1700 cm^{-1} region in the infrared spectra of benzene derivatives. The patterns are useful for identifying substitution type in the benzene ring.

differ from phenanthrene derivatives in that they exhibit an intense band near 1550 cm^{-1}, and the band near 1500 cm^{-1} is absent. Spectra of both tricyclic derivatives contain, in the same region, more skeletal bands than do the spectra of the naphthalene derivatives. For instance, in the Raman spectrum of phenanthrene the bands lie at 1620, 1610, 1560, 1525, 1440 and 1350 cm^{-1}, and in its infrared spectrum they lie at 1620, 1600, 1500 and 1460 cm^{-1} (Table 4.12).

Hetero aromatic compounds with five-membered rings. There are three characteristic skeletal bands in the spectra of five-membered ring compounds. The first appears in the 1610–1570 cm^{-1} region, the second in the 1500–1450 cm^{-1} region and the third in the region 1400–1370 cm^{-1}. All three bands are strong in the infrared spectrum, but in the Raman spectrum only the second and third are intense. Without additional information the five-membered hetero-atom compounds cannot be identified by vibrational spectra. Nevertheless, they can be distinguished from aromatic hydrocarbons, since the ν_{C-H} bands appear at higher frequencies, in the 3150–3100 cm^{-1} region. The ν_{C-H} bands in the infrared spectrum occur in the 800–700 cm^{-1} region.

Table 4.12—Positions of bands corresponding to the skeletal vibration in aromatic hydrocarbons with condensed rings

Compound	Band position, cm^{-1}	Intensity	
		R	IR
Naphthalene[a]	1630–1595	w	m
	1580–1570	vs	vw
	1510–1500	vw	m
	1390–1350	vs	ms
Anthracene[b]	1630–1620	ms	ms
	1560–1550	vs	s
	1400–1390	vs	vw
Phenanthrene[c]	1620–1600	m	m
	1520–1500	m	s
	1460–1440	s	m
	1350–1300	vs	w

[a] Cf. spectrum in Fig. 9.49.
[b] Cf. spectrum in Fig. 9.30.
[c] Cf. spectrum in Fig. 9.32.

Pyridine and its derivatives. The pyridine spectrum (e.g. pyridinium chloride spectrum, Fig. 9.15) is similar to the spectra of monosubstituted benzene derivatives. In the infrared spectrum there are four markedly strong ring skeletal bands, near 1600, 1580, 1500 and 1430 cm^{-1}. Only the first two are intense in the Raman spectrum, in which there is normally a very strong breathing band near 995 cm^{-1}; this band is absent in the spectra

Table 4.13—Bands characteristic of the pyridine ring substitution type*

Substitution type	Vibration	Band position, cm^{-1}
Unsubstituted[a]	$\delta_{C—H}$	1028 IR, R
	$\nu_{skeletal}$	988 R, IR
	$\gamma_{C—H}$	750 IR
	$\delta_{skeletal}$	688 IR
2-Monosubstituted	$\delta_{C—H}$	1050–1040 R
	$\nu_{skeletal}$	1000–995 R
	$\gamma_{C—H}$	780–740 IR
3-Monosubstituted	$\delta_{C—H}$	1040–1030 R
	$\gamma_{C—H}$	820–770 IR
	$\delta_{skeletal}$	730–690 IR
4-Monosubstituted	$\nu_{skeletal}$	995–990 R
	$\gamma_{C—H}$	850–790 IR
2,4-Disubstituted	$\nu_{skeletal}$	1000–995 R
2,5-Disubstituted	$\nu_{skeletal}$	850–840 R
2,6-Disubstituted	$\nu_{skeletal}$	1000–995 R
3,5-Disubstituted	$\delta_{C—H}$	1040–1030 R

* R—strong in the Raman spectrum, IR—strong in the infrared spectrum.
[a] Cf. band in Fig. 9.15.

of 3-, 2,5- and 3,5-pyridine derivatives. The breathing band is often accompanied or replaced by another ($\delta_{C—H}$) in the 1050–1030 cm^{-1} region (Table 4.13). The $\gamma_{C—H}$ bands that are intense in the infrared spectrum can be used to determine the type of substitution, in the same way as for benzene derivatives.

4.2.2 Nitrogen compounds

4.2.2.1 Amines

Primary and secondary amines are characterized by the N–H bands, whereas tertiary amines can be identified only by the $\nu_{C—H}$ bands of the CH_2 and CH_3 groups bonded to nitrogen (cf. Table 4.4, p. 96).

The NH_2 group undergoes two valence vibrations, one of which is symmetric ($\nu_{N—H}$ sym) and appears in the 3450–3300 cm^{-1} region; the other is antisymmetric ($\nu_{N—H}$ as) and appears in the 3550–3300 cm^{-1} region. If one of the N—H groups does not participate in an intramolecular hydrogen bond, the ratio between the symmetric and the antisymmetric frequency is 0.98. In the infrared spectra of aliphatic amines both the $\nu_{N—H}$ bands are of medium strength, and are only fairly intense in the spectra of aromatic amines. In the Raman spectrum the ν_{N-H}sym band is strong but the ν_{N-H}as band is weak. Since the frequencies of the bands are distant from the frequency of the exciting line, an argon laser source should be used to obtain a suitable intensity of these bands (Figs. 9.11 and 9.14). If a helium–neon laser source is used, the bands may be undetectable.

Secondary amines have only one ν_{N-H} band, in the region 3350–3300 cm^{-1} (e.g. morpholine, Fig. 9.12). In the spectra of primary and secondary amines there is normally one weak band near 3200 cm^{-1}, arising from the δ_{N-H} overtone. If the sample is diluted with a non-polar solvent, the weak hydrogen bonds existing between amine molecules will be broken and the ν_{N-H} bands will be shifted to frequencies about 50 cm^{-1} higher.

Table 4.14—Characteristic bands of amines

Group	Type of vibration	Band position, cm^{-1}	Intensity R	Intensity IR
RNH_2[a]	ν_{N-H} as	3550–3350	w	m
	ν_{N-H} sym	3450–3300	s	m
$ArNH_2$[b]	ν_{N-H} as	3500–3400	w	s
	ν_{N-H} sym	3400–3300	s	s
$-NH_2$	δ_{NH_2}	1650–1590	w	ms
	ω_{NH_2}, τ_{NH_2}	850–750	—	s, br
R_2NH[c]	ν_{N-H}	3500–3300	s	m
ArNHR	ν_{N-H}	3500–3300	s	s
	δ_{N-H}	1580–1490	—	w
N—H	ω_{N-H}	750–700	—	s
RCH_2NH_2	ν_{C-N}	1090–1070	m	m
R_2CHNH_2	ν_{C-N}	1150–1080	m	m
R_3CNH_2	ν_{C-N}	1240–1170	m	s
R_2NH	ν_{C-N}	1145–1130	m	m
R_2NH	ν_{C-N} sym	900–850	s	w
$ArNH_2$	ν_{C-N}	1350–1260	s	s
ArNHR	ν_{C-N}	1340–1320	s	s
$ArNR_2$	ν_{C-N}	1380–1310	s	s
R_3N[d]	ν_{C-N} sym	830	s	w
$N-CH_3$	ν_{CH} sym	2810–2770	s	s
	δ_{CH_3} sym	1440–1410	w	m
$N-CH_2$	ν_{CH_2} sym	2820–2760	s	s

[a] Cf. spectrum in Fig. 9.11.
[b] Cf. spectrum in Fig. 9.14.
[c] Cf. spectrum in Fig. 9.12.
[d] Cf. spectrum in Fig. 9.13.

For primary amines the δ_{NH_2} band is of medium strength in both the Raman and infrared spectra (Fig. 9.11), occurring in the 1650–1590 cm^{-1} region. The δ_{N-H} band is generally undetectable in the spectra of secondary amines. The ω_{NH_2} and τ_{NH_2} vibrations tend to give rise to broad and fairly intense bands in the 850–700 cm^{-1} region, but only in the infrared spectrum. The ν_{C-N} bands are less characteristic of aliphatic amines but are strong in the spectra of aromatic amines. They can be used to indicate the presence of tertiary amines. The symmetric stretching vibration of the C—N—C fragment gives rise to a strong band, which can be useful for identification. This band appears in the Raman spectra of secondary amines in the 900–850 cm^{-1} region and in the spectra of tertiary amines, usually near 830 cm^{-1} (Table 4.14). Tertiary amines, as mentioned above, can be identified by the position of the symmetric ν_{C-H} bands of the methyl and methylene groups adjacent to the nitrogen atom, since these bands appear at lower frequencies than the corresponding bands of alkanes (Fig. 9.13). Another method of identifying tertiary amines is to convert the amines into their salts, which have broad intense bands in the 2700–2300 cm^{-1} region of their infrared spectra.

4.2.2.2 Amine salts

Raman spectra are not useful for identifying amine salts, because bands characteristic of these compounds appear only in the infrared spectrum. In the spectra of the salts of primary and secondary amines there are some bands (due to composite tones) in the 2700–2000 cm^{-1} region, which appear in addition to a broad band in the 3200–2800 cm^{-1} region. One of these bands can usually be observed near 2000 cm^{-1}. Tertiary amine salts show intense absorption in the 2700–2300 cm^{-1} region. The position of the bands, characteristic of the substituted ammonium group, depends, to a certain degree, on the anion. Most typical are the spectra of hydrochlorides. Their characteristic bands are given in Table 4.15 (cf. also the spectrum in Fig. 9.15).

4.2.2.3 Imines (the $>$C=N—H *group)*

Schiff's bases, hydrazones, semicarbazones and oximes are characterized by the C=N stretching band. This band appears in the region 1690–1610 cm^{-1}, as does the $\nu_{C=C}$ band. The former is usually stronger than the latter in the infrared spectrum but slightly weaker in the Raman spectrum (e.g.) the oxime spectrum in Fig. 9.16), so the two groupings (C=C, C=N

Table 4.15—Characteristic infrared bands of amine salts

Group	Type of vibration	Band position, cm^{-1}	Intensity
$—NH_3^+$	ν_{NH_3} as and sym overtones and combination tones	3200–2800	vs, br
		2800–2000	m
		2000	m
	δ_{NH_3} as	1620–1570	m
	δ_{NH_3} sym	1550–1500	m
$—NH_2^+$	ν_{NH_2} as and sym overtones and combination tones	3000–2700	vs, br
		2700–2300	m
	δ_{NH_2}	1600–1570	m
$—NH^+$	$\nu_{N—H}$, overtones and combination tones	2700–2300	vs, br

can normally be distinguished by spectroscopic methods. The infrared spectra of oximes contain a very strong $\nu_{O—H}$ band (weak in the Raman spectrum) in the 3300–3150 cm^{-1} region, and a band near 950 cm^{-1} (Fig. 9.16) ascribed to the $\nu_{N—O}$ vibration. Spectra of hydrazones contain bands characteristic of the NH_2 group ($\nu_{N—H}$), whereas in semicarbazone spectra bands appear that are characteristic of primary amides (Table 4.16).

Table 4.16—Positions of the $\nu_{C=N}$ band*

Group	Band position, cm^{-1}
RCH=NR	1675–1665
ArCH=NR	1655–1630
ArCH=NAr	1640–1625
R_2C=NR	1660–1650
ArRC=NR	1650–1640
ArR=CNAr	1640–1630
RCH=NOH[a]	1675–1650
ArCH=NOH	1645–1615
R_2C=NOH	1685–1655
ArRC=NOH	1640–1620
R_2C=N—NH_2	1650–1610
R_2C=N—NHAr	1640–1620
R_2C=$NNHCONH_2$	1665–1650
ArRC=$NNHCONH_2$	1620–1610

* The band is strong in both Raman and infrared spectra.
[a] Cf. spectrum in Fig. 9.16.

4.2.2.4 Nitriles

The —C≡N group is characterized by a band arising from the $\nu_{C \equiv N}$ stretching vibration. This band appears in the 2260–2240 cm^{-1} region, as do bands due to the stretching vibration of the carbon–carbon triple bonds of disubstituted acetylenes. The former band is usually stronger in the infrared than in the Raman spectrum, and the latter band is absent from the infrared spectrum (for symmetric compounds), or very weak. If the cyano group forms part of a conjugated system, the $\nu_{C \equiv N}$ frequency will be about 20 cm^{-1} lower (Table 4.17). The intensity of the $\nu_{C \equiv N}$ band in

Table 4.17—Positions of the $\nu_{C \equiv N}$ band in the spectra of nitriles

Group	Band position, cm^{-1}	Intensity	
		R	IR
R—C≡N[a]	2260–2240	vs	m
C=C—C≡N	2235–2215	vs	s
Ar—C≡N[b]	2240–2220	vs	s

[a] Cf. spectrum in Fig. 9.17.
[b] Cf. spectrum in Fig. 9.18.

the Raman spectra of aromatic nitriles is much higher than that in the spectra of aliphatic nitriles (cf. the spectra in Figs. 9.17 and 9.18). The Raman spectra of aromatic nitriles contain a strong band arising from the δ_{C-CN} bending vibration in the 400–350 cm^{-1} region, whereas the spectra of aliphatic nitriles are characterized by a band in the 200–160 cm^{-1} region (skeletal deformation vibration).

4.2.2.5 Azo compounds

Aliphatic azo compounds are characterized by a band near 1580 cm^{-1} in the Raman spectrum (Fig. 9.17). This band is absent in the infrared spectra of azo compounds in the *trans*-form (according to the selection rules). Aromatic azo compounds are characterized by a very strong Raman band in the 1430–1400 cm^{-1} region. This band is also visible in the infrared spectrum if the —N=N— group is substituted asymmetrically (Fig. 9.19).

4.2.2.6 Azides

Organic azides are characterized by a band in the 2170–2080 cm^{-1} region, arising from the antisymmetric vibration of the ($-\overline{N}=\overset{\oplus}{N}=\overset{\ominus}{\underline{N}}| \leftrightarrow -\overset{\ominus}{\underline{N}}-\overset{\oplus}{N}\equiv\overline{N}$) group. This band is strong in the infrared spectrum and weak in the Raman spectrum. The symmetric vibration gives rise to a strong Raman band in the 1340–1180 cm^{-1} region (the phenyl azide spectrum shown in Fig. 9.20 may be used as an example). In the spectra of acyl azides there is usually a doublet (ν_{N_3} as) in the 2240–2180 cm^{-1} and 2160–2140 cm^{-1} regions as well as a band (ν_{N_3} sym) in the region 1260–1240 cm^{-1}.

4.2.2.7 Nitroso compounds

Aliphatic *C*-nitroso compounds absorb in the region 1600–1550 cm^{-1}, whereas the aromatic compounds absorb in the 1540–1490 cm^{-1} region. The $\nu_{N=O}$ bands are of medium intensity in the infrared spectrum. The Raman spectra of nitroso compounds are difficult to obtain because of their deep blue or green colour.

In *N*-nitrosamines the —N=O band appears in the region 1500–1430 cm^{-1} (Fig. 9.21). This band, being of medium intensity, can easily be confused with the δ_{C-H} deformation band. The $\nu_{N=O}$ bands appear in the 1650–1640 cm^{-1} region in the spectra of organic nitrites.

4.2.2.8 Nitro compounds

The characteristic vibrations of the nitro group are weakly coupled with the remaining vibrations of the molecule, so the presence of a nitro group can easily be deduced from the spectrum. Both N—O bonds in the nitro group, being of one-and-a-half order, are equivalent. The coupling of these two identical oscillators causes the nitro group to undergo an antisymmetric stretching vibration as well as a symmetric stretching vibration, resulting in the appearance of two bands, one in the 1560–1500 cm^{-1} region (ν_{NO_2} as), the other in the 1380–1330 cm^{-1} region (ν_{NO_2} sym). Both bands are very intense in the infrared spectrum but the antisymmetric is always stronger than the symmetric band. In the Raman spectra, the symmetric band of aliphatic nitro compounds is of medium strength (Fig. 9.22), whereas the corresponding band of aromatic compounds is extremely intense (Fig. 9.23). The antisymmetric vibration tends to give rise to a weak or very weak band. Electronegative substituents (for example halogens) increase the antisymmetric frequency and decrease the symmetric frequency. An aromatic ring or a C=C group lowers both frequencies. If the nitro group is hydrogen bonded the symmetric frequency will be considerably lowered (even by as much as 100 cm^{-1} in, for example, 2-nitroaniline). The nitro group lowers the deformation frequency of a neighbouring methylene group (1440–1410 cm^{-1}). If, however, instead of the methylene group, there is a methyl group in close proximity to the nitro group, the symmetric vibrations δ_{NO_2} and δ_{CH_3} couple, resulting in two strong bands near 1400 and 1360 cm^{-1}, due to the CH_3—C—NO_2 grouping (Fig. 9.22). The presence of a nitro group in aromatic compounds changes the $\gamma_{C—H}$ frequency (usually increasing it), thus making it difficult to identify the type of substitution in the benzene ring by the position of these bands in the infrared spectrum. In addition to the N—O stretching bands, the nitro group is characterized by some other bands arising from two further types of vibration. The C—N stretching vibration ($\nu_{C—N}$) gives rise to a strong Raman band in the 900–800 cm^{-1} region; this band is of medium strength in the infrared spectrum. The deformation vibration (δ_{NO_2}) gives rise to a medium band in the 700–650 cm^{-1} region in both types of spectra (Table 4.18).

Table 4.18—Bands characteristic of the nitro group

Group	Type of vibration	Band position, cm^{-1}	Intensity R	Intensity IR
R—NO_2[a]	ν_{NO_2} as	1560–1545	vw	vs
	ν_{NO_2} sym	1390–1355	m	vs
C=C—NO_2	ν_{NO_2} as	1550–1500	w	vs
	ν_{NO_2} sym	1360–1290	s	vs
Ar—NO_2[b]	ν_{NO_2} as	1530–1500	w	vs
	ν_{NO_2} sym	1370–1330	vs	vs
R—NO_2	$\nu_{C—N}$	920–830	vs	s
Ar—NO_2	$\nu_{C—N}$	850–750	s	s
$CHXNO_2$	ν_{NO_2} as	1580–1560	—	vs
(X=Cl, Br)	ν_{NO_2} sym	1370–1340	m	vs
N—NO_2	ν_{NO_2} as	1630–1570	m	vs
	ν_{NO_2} sym	1315–1260	m	vs
O—NO_2	ν_{NO_2} as	1660–1625	—	vs
	ν_{NO_2} sym	1285–1270	m	vs
—NO_2	δ_{NO_2}	700–650	m	m
	γ_{NO_2}	620–610	w	m
	ϱ_{NO_2}	560–480	m	m

[a] Cf. spectrum in Fig. 9.22.
[b] Cf. spectrum in Fig. 9.23.

4.2.3 Oxygen compounds

4.2.3.1 Alcohols and phenols

Alcohols and phenols are generally characterized by the O—H stretching vibration. In spectra of pure substances it gives rise to a broad band in the

3400–3200 cm^{-1} region, because phenols and alcohols tend to form strong hydrogen bonds. The band is very strong in the infrared spectrum but weak (or even very weak in the spectra of phenols) in the Raman spectrum, so an argon-laser exciting source is necessary for its detection. The ν_{N-H} bands may also appear in approximately the same region and are markedly stronger in the Raman spectrum. Alcohols and phenols, diluted with non-polar solvents (CCl_4, C_6H_{14}), in which the hydroxyl compounds do not tend to form intermolecular hydrogen bonds, exhibit a sharp band near 3600 cm^{-1}, characteristic of the free hydroxyl group. In practice, however, this band is observable only in the infrared spectrum. Its position supplies information on the alcohol type (primary, secondary, tertiary) and allows alcohols to be distinguished from phenols. In spectra of alcohols or phenols which form intramolecular hydrogen bonds, the ν_{O-H} band is not shifted to higher frequencies when the sample is diluted with a non-polar solvent; also, the stronger the hydrogen bond, the lower the ν_{O-H} frequency. Intramolecular hydrogen bonds exist in phenols which contain, in the *ortho* position, groups having proton acceptor properties, such as —C=O (Fig. 9.41), —NO_2, —NR_2, or a halogen atom (Fig. 9.25). The carbonyl group tends to form the strongest hydrogen bonds and the halogens form the weakest. If the intramolecular hydrogen bond is very strong, as, for example, in the keto–enol forms of chelating agents, the ν_{O-H} band becomes very broad and its intensity decreases, so the band may be undetectable in the spectrum. The O—H bending vibration is of little use for identification purposes. It gives rise to broad bands which are of medium strength in the infrared spectrum and weak in the Raman spectrum. The bands lie in the 1420–1380 cm^{-1} region in the spectra of alcohols, in the 1390–1330 cm^{-1} region in the spectra of phenols (δ_{O-H}), and in the 760–660 cm^{-1} region for alcohols and phenols (γ_{O-H}). In addition to bands characteristic of the hydroxyl group, the spectra of phenols also contain bands characteristic of the aromatic ring.

Bands due to the C—O skeletal stretch are also useful for diagnostic purposes. One of these bands, antisymmetric in character, is strong in the infrared spectrum and of medium intensity in the Raman spectrum. Its position depends on the type of alcohol. In the spectra of primary alcohols this band appears near 1060 cm^{-1}, and in the spectra of secondary and tertiary alcohols near 1100 and 1200 cm^{-1}, respectively. In phenols the band lies in the 1260–1200 cm^{-1} region. The ν_{C-O} frequency in cyclic alcohols depends on the position of the hydroxyl group. For the equatorial OH group the frequency falls in the region 1065–1040 cm^{-1}, for the axial

Table 4.19—Bands characteristic of alcohols and phenols

Group	ν_{O-H} band position, cm^{-1}	Intensity R	Intensity IR	Group	Type of vibration	Band position, cm^{-1}	Intensity R	Intensity IR
RCH_2OH	3640		s	RCH_2OH[a]	ν_{C-O}[b]	1075–1050	s	vs
R_2CHOH	3630		s		ν_{CCO} sym	970–960 [c]	vs	
R_3COH	3620		s	R_2CHOH	ν_{C-O}	1125–1090	s	vs
ArOH	3610		s		ν_{CCO} sym	830–810	vs	
—OH...O (intermolecular)	3400–3200	w	vs,br	R_3COH	ν_{C-O}	1210–1120	m	vs
					ν_{CCO} sym	760–730	s	
—OH...X (intramolecular)	3600–2600	w	va,br	cyclic[d]				
				equatorial OH group	ν_{C-O}	1065–1040	m	vs
				axial OH group	ν_{C-O}	1030– 970	m	vs
				ArOH[e]	ν_{C-O}	1260–1180	m	vs

[a] Cf. spectrum in Fig. 9.11.
[b] The ν_{C-O} band position given is for strongly associated molecules.
[c] For CH_3OH the frequency is 1030, for C_2H_5OH it is 885 cm^{-1}.
[d] Cf. spectrum in Fig. 9.24.
[e] Cf. spectra in Figs. 9.19, 9.25, 9.26, 9.41.

OH group it appears in the range 1035–970 cm^{-1} (cf. the cyclopentanol spectrum in Fig. 9.24). The symmetric skeletal vibration gives rise to strong Raman bands in the spectra of primary, secondary and tertiary alcohols, in the regions 970–960, 830–810 and 770–700 cm^{-1}, respectively (Table 4.19).

4.2.3.2 Ethers

The spectra of ethers do not contain characteristic bands which are easy to identify. The C—O—C grouping can vibrate symmetrically and antisymmetrically. Since the atomic mass of carbon differs only slightly from that of oxygen, and the two force constants are close in value, both modes are strongly coupled with the remaining skeletal vibration. Moreover, the ν_{C-O} bands lie in the 1200–800 cm^{-1} region, in which numerous other skeletal bands may appear. Nevertheless, if a molecule containing the ether group does not contain other functional groups, the presence of an ether can be spectroscopically confirmed by the strong C—O—C antisymmetric band in the 1200–1100 cm^{-1} region in the infrared spectrum, and the symmetric band in the 930–830 cm^{-1} region in the Raman spectrum. The spectrum of di-(3-methylbutyl) ether may be given here as an example (Fig. 9.27). If the chain is branched in the α-position, the symmetric band frequency decreases to 820–800 cm^{-1}. Further chain branching (tert-butyl group) causes the symmetric band frequency to decrease even further, to 770–700 cm^{-1}. Diaryl ethers are characterized by an infrared band near 1240 cm^{-1}, whereas alkyl aryl ethers have two infrared bands, near 1250 cm^{-1} (Figs. 9.14 and 9.33) and 1020 cm^{-1}, the latter (ν_{C-O} sym) being also intense in the Raman spectrum (Table 4.20). Cyclic ethers are characterized by bands arising from ring vibrations. In particular, the position of the ring breathing band, which is strong in the Raman spectrum, is a good indication of ring size (cf. the spectrum of 1-chloro-2,3-epoxypropane in Fig. 9.28, and Table 4.1, p. 87).

In the spectra of acetals there are four bands in the 1190–1030 cm^{-1} region. They are fairly intense in the infrared spectrum and of medium strength in the Raman spectrum. The symmetric vibration of the acetal grouping (C—O—C—O—C) gives rise to a strong Raman band in the 900–800 cm^{-1} region.

The presence of the ether group can be additionally confirmed by the band positions of the CH_2 and CH_3 groups bonded to an oxygen atom (cf. Section 4.2.1.3).

Table 4.20—Positions of the characteristic ν_{C-O} band in the spectra of ethers

Group	Band position, cm^{-1}	Intensity R	Intensity IR
R—O—R[a]	1150–1070	m	vs
	930–830[b]	vs	m
R—O—C=C	1225–1220	m	vs
	1120–1030	m	s
R—O—Ar[c]	1300–1210	m	vs
	1030–1010	m	vs
Ar—O—Ar	1270–1230	m	vs
$R_2C(OR)_2$	1190–1030[d]	m	vs
	900–800	vs	
O[e] C—C	1270	vs	s
	870	m	s
C C O C	990–970	w	vs
	1030	vs	m
C C_2 O C	1090–1070	w	vs
	950–900	vs	m
Dioxane	1120	w	vs
	835	vs	

[a] Cf. spectrum in Fig. 9.27.

[b] A simple branching of the chain in the α-position (e.g. the isopropyl group) lowers the frequency to near 800 cm^{-1}, a double branching (e.g. the tert-butyl group), to near 700 cm^{-1}.

[c] Cf. spectra in Figs. 9.14, 9.33.

[d] Four bands.

[e] Cf. spectrum in Fig. 9.28.

4.2.3.3 Carbonyl compounds

All carbonyl compounds are characterized by the $\nu_{C=O}$ band, which is very strong in the infrared spectrum and of medium strength in the Raman spectrum. This band lies in the 1900–1600 cm^{-1} region. The identification of a particular type of carbonyl compound consists in determining the $\nu_{C=O}$ band frequency and confirming the presence of other functional groups in the molecule examined. The frequency of the C=O stretching vibration depends on the angles between neighbouring bonds (Fig. 9.31), the inductive and mesomeric effects of neighbouring groups (Figs. 9.29–9.45) and, in the case of unsaturated compounds, on the coupling of the C=O and C=C oscillators (Fig. 9.44). If the carbonyl group participates in a hydrogen bond, the $\nu_{C=O}$ frequency decreases markedly (even by as much as 100 cm^{-1}, cf. Fig. 9.41). All these effects have been discussed previously (see Section 4.1.3.2). If the carbonyl group is conjugated with a carbon–carbon double bond, or with an aromatic ring, the $\nu_{C=O}$ frequency (characteristic of the given carbonyl compound type) will be lowered by about 20–30 cm^{-1}. The carbonyl group also has a characteristic effect on the vibrational frequencies of the CH_2 and CH_3 groups bonded to C=O, increasing their valence frequencies (ν) and decreasing the deformation frequencies (δ) (see Section 4.2.1.3).

Aldehydes. The spectra of aldehydes differ from those of other carbonyl compounds, in that they contain, in addition to a $\nu_{C=O}$ band, a characteristic doublet in the 2850–2700 cm^{-1} region. The doublet is due to Fermi resonance of the carbon–hydrogen stretching vibration of the aldehyde group (—CHO) with the carbon–hydrogen bending vibration overtone (ϱ_{C-H}). Both bands of the doublet are of medium strength in the infrared and Raman spectra. The band of higher frequency may sometimes overlap the ν_{C-H} bands of aliphatic groups, but the band of lower frequency is always clearly visible in the spectrum. In the spectra of aliphatic aldehydes the $\nu_{C=O}$ band appears in the 1740–1720 cm^{-1} region. Its frequency in unsaturated aldehydes is 20–40 cm^{-1} lower. In the spectra of α-haloaldehydes the $\nu_{C=O}$ band does not exhibit splitting. Aldehydes can also be characterized by the ϱ_{C-H} band, near 1390 cm^{-1}, which is easily observed in the Raman spectrum (Table 4.21). In the spectra of aromatic aldehydes there is, moreover, an intense band due to the skeletal vibration, which appears in the 1250–1160 cm^{-1} region (Fig. 9.30). The band occurs at lower frequencies in the spectra of aliphatic aldehydes. Its position in the spectrum is also influenced by chain branching in the α-position (cf. the spectrum of α-methylbutanal, Fig. 9.29).

Ketones. In non-cyclic saturated ketones the carbonyl group gives rise to a band in the 1725–1705 cm^{-1} region. This band appears in the spectra of unsaturated ketones in the 1700–1600 cm^{-1} region, i.e. at about 10–20 cm^{-1} lower than the corresponding band of aldehydes of similar structure. Halogens in the α-position increase the $\nu_{C=O}$ frequency by up to 25 cm^{-1}, depending on the steric arrangement of the C—Cl and C=O bonds. The maximum frequency shift may be observed when the bonds are synperiplanar, the minimum when they are synclinal. Each conformation gives rise to a separate band. Nevertheless, the presence of two bands in a spectrum can be associated with several different stable conformations, each of them having a slightly different spectrum (for example α-haloketones, vinyl ketones, or *ortho* derivatives of acetophenone). The splitting of bands may also result from Fermi resonance of the $\nu_{C=O}$ vibration with an overtone of another vibration. The $\nu_{C=O}$ frequency in cyclic ketones with six-membered (or larger) rings has the same value as that for non-

Table 4.21—Bands characteristic of aldehydes

Group	Type of vibration	Band position, cm^{-1}	Intensity R	Intensity IR
R—CHO[a]	$\nu_{C=O}$	1740–1720	s	vs
C=C—CHO	$\nu_{C=O}$	1700–1680	s	vs
$(C=C)_2$—CHO	$\nu_{C=O}$	1680–1660	s	vs
Ar—CHO[b]	$\nu_{C=O}$	1710–1690	s	vs
—CHO	ν_{C-H}	2830–2810	m	m
	$2\varrho_{C-H}$	2720–2700	m	m
—CHO	ϱ_{C-H}	1400–1380	s	m
ArCHO	$\nu_{skeletal}$	1230–1160	s	vs
RCHO	$\nu_{skeletal}$	1120–1090	s	
R_2CCHO	$\nu_{skeletal}$	790–750	s	
—CHO	$\delta_{C=O}$	540–510	m	

[a] Cf. spectrum in Fig. 9.29.
[b] Cf. spectrum in Fig. 9.30.

cyclic ketones. This frequency is about 20 cm^{-1} higher for cyclopentanone derivatives and about 60 cm^{-1} higher for cyclobutanone derivatives (Fig. 9.31). If the carbonyl group forms a hydrogen bond, its frequency tends to decrease. The frequency shift is particularly large in the case of chelates. 2-Hydroxyacetophenones, for example, absorb in the 1670–1640 cm^{-1} region, whereas *cis*-keto–enols absorb in the 1640–1580 cm^{-1} region (Table 4.22). In the spectra of aliphatic ketones there is an additional band in the 1275–1025 cm^{-1} region, arising from the skeletal vibration. It is strong in the infrared spectrum and of medium strength in the Raman spectrum. The corresponding band of aromatic ketones appears in the 1325–1215 cm^{-1} region. The C=O deformation vibration gives rise to a band of medium intensity in both spectra, in the 600–550 cm^{-1} region.

In the spectra of diazoketones the $\nu_{C=O}$ band appears in the 1650–1640 cm^{-1} region for aliphatic compounds, and in the 1630–1605 cm^{-1} region for aromatic compounds. The low value of the $\nu_{C=O}$ frequency is due to the decreasing carbon–oxygen bond order and the increasing contribution of canonical structures of the type $\overset{\ominus}{O}—C{=}CH—\overset{\oplus}{N}{\equiv}N$. Moreover, the diazo group gives rise to a band in the 2100–2080 cm^{-1} (νas) and 1390–1330 cm^{-1} (νsym) regions. All three bands are strong in the infrared spectrum but the third band is weak in the Raman spectrum (Fig. 9.33).

Table 4.22—Positions of the $\nu_{C=O}$ band in the spectra of ketones*

Group	Band position, cm^{-1}
RCOR	1725–1705
tert-Bu_2CO	1685
RCOC=C	1700–1670
$(C{=}C)_2CO$	1680–1640
ArCOR	1700–1680
ArCOAr	1670–1660
Cyclanones:	
six-membered or larger ring	1725–1705
five-membered ring	1750–1740
four-membered ring[a]	1780
Tropolone	1620
o-HOArCOR	1670–1630
cis-Keto-enols	1640–1580
RCOCOR	1730–1710
ArCOCOAr	1680
RCOCHCl	1745
	1725
$RCOCCl_2$	1755
Quinones[b]	1680–1640

* The band is very strong in the infrared but medium or weak in the Raman spectrum.
[a] Cf. band in Fig. 9.31.
[b] Cf. spectra in Figs. 9.32, 9.33.

Carboxylic acids. Carboxylic acids are strongly associated and exist generally as dimers, even in solution in non-polar solvents:

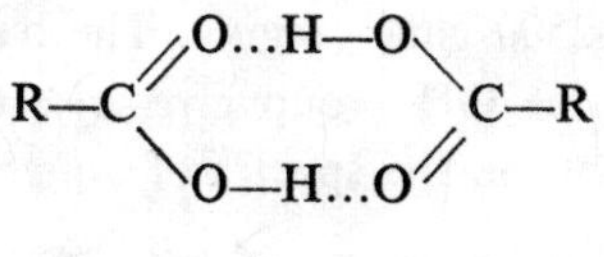

```
        O...H—O
       //       \
R—C              C—R
       \        //
        O—H...O
```

Catenated species may also occur:

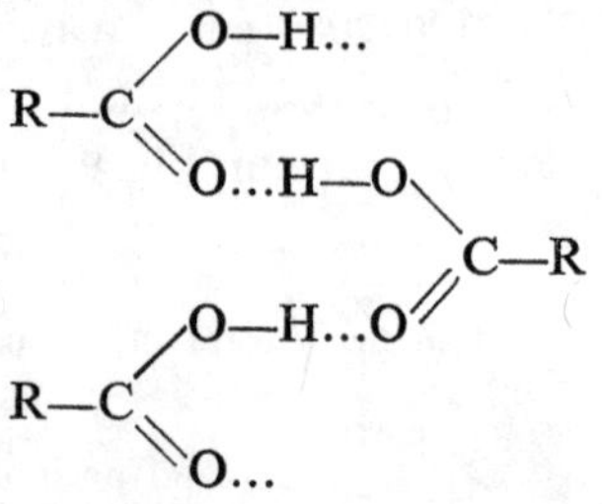

```
         O—H...
        /
   R—C
        \\
         O...H—O
                \
                 C—R
                //
         O—H...O
        /
   R—C
        \\
         O...
```

In a dimer there are two carbonyl groups, the coupled valence vibrations of which give rise to one symmetric and one antisymmetric vibration. Since the dimer has a centre of symmetry, the symmetric vibration will be Raman-active, whereas the antisymmetric mode will be infrared-active. The symmetric vibration gives rise to a strong Raman band in the 1670–1630 cm^{-1} region, and the antisymmetric to a very strong infrared band in the 1730–1680 cm^{-1} region (Fig. 9.34). This lack of correlation between bands in the Raman and infrared spectra is a characteristic feature of carboxylic acid dimers. Dicarboxylic acids generally exist as linear polymers. Again, the $\nu_{C=O}$ band position in the Raman spectrum differs from that in the infrared spectrum (Fig. 9.35). In the monomer, which exists in small quantities in a non-polar environment, the $\nu_{C=O}$ band appears in the region 1760–1740 cm^{-1} (in the same position in both spectra). In solvents which can act as acceptors in hydrogen bonding (alcohols, ethers, e.g. dioxane), carboxylic acids exist as monomers, hydrogen-bonded to the solvent molecules. In this case, the $\nu_{C=O}$ vibration gives rise to a band in the 1730–1710 cm^{-1} region. The influence of the structure of the remaining part of the molecule on the $\nu_{C=O}$ vibration frequency of the acid is similar to that exerted on the $\nu_{C=O}$ frequencies of other carbonyl compounds. Halogens in the α-position increase the $\nu_{C=O}$ frequency; a carbon–carbon double bond or an aromatic ring lowers this frequency. Strong intramolecular hydrogen bonds in α-hydroxy or α-amino aromatic acids give rise to an additional decrease in the $\nu_{C=O}$ frequency.

Associated acids are characterized moreover by a strong broad infrared band, attributed to the ν_{O-H} vibration, in the 3200–2500 cm^{-1} region. This broad band often exhibits small sharp peaks (corresponding to the dimer composite tones) in the 2700–2500 cm^{-1} region. The band is very weak in the Raman spectrum. The free OH group gives rise to a sharp band in the 3580–3500 cm^{-1} region, but in the spectra of dimers there are

Table 4.23—Bands characteristic of carboxylic acids

Group	Position of the $\nu_{C=O}$ band,* cm^{-1} R	IR	Type of vibration of the COOH group	Band position, cm^{-1}	Intensity R	IR
RCOOH[a, b]	1680–1650	1725–1700	ν_{O-H}	3200–2500	w	s,vbr
RCHXCOOH	1700	1740–1720	overtones	2700–2500	w	m
RCX_2COOH	1680	1750	δ_{O-H}	1440–1395	m	m
CCl_3COOH	1690	1745	ν_{C-O}	1300–1250	w	s
CF_3COOH	1770	1785	ν_{O-H}	950–900	—	m,br
C=CCOOH[c]	1690–1630	1715–1690	$\delta_{C=O}$	640–620	sm	sm
ArCOOH	1680–1620	1700–1690				
ArCOOH			Monomer in dioxane			
(with intramolecular H—bond)	1640–1620	1670–1650	ν_{O-H}	3580–3500	w	m
			$\nu_{C=O}$	1760–1740	m	s
			$\nu_{C=O}$	1735–1715	m	vs

* The band is medium in the Raman spectrum but very strong in the infrared spectrum.

[a] The band positions given refer to associated acids (mostly cyclic dimers). If carboxylic acids form intramolecular bonds, the band positions in both spectra are identical.

[b] Cf. spectrum in Fig. 9.34.

[c] Cf. spectrum in Fig. 9.35.

three bands characteristic of the carboxyl group. Two of them are attributed to the coupled vibrations δ_{O-H} (1440–1395 cm^{-1}) and ν_{C-O} (1300–1250 cm^{-1}); the third, which is fairly broad and visible only in the infrared spectrum, arises from the γ_{O-H} vibration (950–880 cm^{-1}). The band attributed to the $\delta_{C-C=O}$ deformation mode appears in the 640–620 cm^{-1} region; it is usually strong in both spectra (Table 4.23).

Both carbon–oxygen bonds in the carboxylate group, being of one-and-a-half order, are therefore equivalent. Thus, this group is characterized by two valence vibrations, one antisymmetric and at higher frequency, and the other symmetric, at lower frequency. The antisymmetric band, which lies in the 1620–1550 cm^{-1} region, is weak in the Raman spectrum but very strong in the infrared spectrum. The symmetric band, which falls in the region 1440–1360 cm^{-1}, is strong in both spectra.

Amino-acids. Amino-acids generally exist as zwitterions $RCH(\overset{+}{N}H_3)COO^-$, so they have bands characteristic of both the ammonium and carboxylate groups (cf. the L-cystine spectrum in Fig. 9.36). These bands are strong in the infrared spectrum but weak or of medium intensity in the Raman spectrum. The $\overset{+}{N}H_3$ group gives rise to a broad band in the region 3100–2700 cm^{-1} (ν_{N-H} as and ν_{N-H} sym), which is accompanied in the 2700–2000 cm^{-1} region by weak bands arising from composite tones and two other bands attributed to the deformation vibrations (δ_{NH_3}), in the 1660–1480 cm^{-1} region. Bands assigned to the carboxylate group appear in the 1610–1560 cm^{-1} region (ν_{C-O} as), and the 1420–1395 cm^{-1} region (ν_{C-O} sym). The latter is markedly strong only in the Raman spectrum. In the spectra of amino-acid hydrochlorides the carbonyl $\nu_{C=O}$ band, characteristic of carboxylic acids, appears in the 1750–1700 cm^{-1} region in the infrared spectrum and in the 1680–1640 cm^{-1} region in the Raman spectrum. In Raman spectra of amino-acids the ν_{C-H} bands are clearly visible but these bands are weak or overlaid by the ν_{N-H} bands in the infrared spectra (Table 4.24 and Fig. 9.36).

Amides. The characteristic $\nu_{C=O}$ band occurs at a particularly low frequency in the spectra of amides, which reflects the mesomeric structure of the amide group:

$$\underset{\underset{O}{\|}}{-C}-NR_2 \leftrightarrow \underset{\underset{O^-}{|}}{-C}=\overset{+}{N}R_2$$

Table 4.24—Bands characteristic of amino-acids*

NH_3^+ group				COO^- group			
Type of vibration	Band position, cm^{-1}	Intensity R	Intensity IR	Type of vibration	Band position, cm^{-1}	Intensity R	Intensity IR
ν_{N-H}	3100–3000	w	vs, br	ν_{CO} as	1610–1560	wm	vs
δ_{NH_3} as	1660–1590	mw	m	ν_{CO} sym	1420–1395	s	s
δ_{NH_3} sym	1530–1490	m	s				

* Cf. spectrum in Fig. 9.36.

Bond equalization within the amide group causes the valence vibrations to lose, to a marked extent, their localization. The C—O and C—N stretching vibrations in primary and secondary amides are accompanied by the N—H bending motion. The complex character of amido group vibrations was the reason for applying a special notation for bands attributed to specific vibrations. The band corresponding to the vibration having the greatest contribution of the $\nu_{C=O}$ mode, which appears at the highest frequency (in the 1680–1630 cm^{-1} range) is identified as Amide I; the band in which the δ_{N-H} mode contributes markedly is called Amide II and the band with an important contribution of the ν_{C-N} vibration is termed Amide III. The Amide I band is strong in the infrared spectrum and of medium intensity in the Raman spectrum. In the spectra of tertiary amides there is a characteristic Amide III band of medium strength near 1500 cm^{-1}. Strong Raman bands at 870–820 cm^{-1} are exhibited by formamides and at 750–680 cm^{-1} by other amides. The latter band is attributed to the symmetric valence vibration of the C—N—C group. The Amide II band appears near 1610 cm^{-1} in the spectra of primary amides, and near 1550 cm^{-1} in the spectra of secondary amides (Figs. 9.37 and 9.38). The band is very strong in the infrared spectrum but weak or absent in the Raman spectrum. The Amide III band appears near 1400 cm^{-1} in the spectra of primary amides, and near 1200 cm^{-1} in those of secondary amides. It is strong in both the infrared and Raman spectra. In primary amides the NH_2 group is, additionally, characterized by two bands arising from the N—H valence vibration; one near 3350 cm^{-1} (ν_{N-H} as) the other near 3180 cm^{-1} (ν_{N-H} sym). Often, however, four bands appear in the spectra of pure samples, as a result of the formation of various aggregates (cyclic dimers, chain aggregates). These bands fall near 3500, 3350, 3300 and 3180 cm^{-1} (cf. the spectrum of formamide in Fig. 9.37). In secondary amides the ν_{N-H} band lies near 3300 cm^{-1}. It is accompanied by a band attributed to an overtone of the Amide II frequency (Fig. 9.38). Bands due to the ν_{N-H} vibration are strong in the infrared spectrum but of medium strength in the Raman spectrum. The strong association exhibited by primary and secondary amides results in considerable frequency shifts when samples are diluted with non-polar solvents.

Cyclic amides (lactams) are characterized by the Amide I band (Fig. 9.39). The Amide II band is absent in their spectra and is also absent in the spectra of cyclic imides. Spectra of non-cyclic imides are similar to those of secondary amides. In the spectra of cyclic imides there are two Amide I bands (Table 4.25).

Table 4.25—Bands characteristic of amides

Type of vibration	Band position, cm^{-1} Associated molecules	Band position, cm^{-1} Non-associated molecules	Intensity R	Intensity IR
Primary amides:[a]				
ν_{NH_2} as	sh. 3350	3500	m	s
ν_{NH_2} sym	sh. 3180	3400	m	s
Amide I (ν_{C-O})	1650±20	1685±15	m	vs
Amide II (ν_{N-H})	1640–1620	1620–1585	w	s
Amide III (ν_{C-N})	1430–1390	1430–1390	ms	m
ϱ_{NH_2}	1150–1100			m
δ_{OCN}	600–550		s	s
Secondary amides:[b]				
ν_{N-H}	3320–3270	3460–3400	m	s
Amide II overtone	sh. 3100	3100	w	m
Amide I	1680–1630	1700–1660	m	vs
Amide II	1570–1510	1550–1510	vw	vs
Amide III	1300–1250	1250–1200	s	s
γ_{N-H}	750–700		w	s
δ_{OCN}	620		s	s
Tertiary amides:[c]				
Amide I ($\nu_{C=O}$)	1650±20		m	vs
Amide III (ν_{C-N})	sh. 1500		m	m
ν_{CNC} $HCONR_2$	870–820		vs	—
ν_{CNC} $RCONR_2$	750–700		vs	—
δ_{OCN}	650–600		s	s
Lactams:				
ν_{N-H}	sh. 3200		m	s
Overtone	sh. 3100		w	m
Amide I { six-membered	1650		m	vs
Amide I { five-membered[c]	1725±25		m	vs
Amide I { four-membered	1745±15		m	vs

[a] Cf. spectrum in Fig. 9.37.
[b] Cf. spectrum in Fig. 9.38.
[c] Cf. spectrum in Fig. 9.39.

Esters. Aliphatic esters are characterized by the $\nu_{C=O}$ band in the 1750–1735 cm^{-1} region. Its frequency is about 20 cm^{-1} lower in the spectra of formates, and about 20–30 cm^{-1} higher in esters of phenols (or enols). Halogens in the α-position increase the $\nu_{C=O}$ frequency by 10–40 cm^{-1} (Fig. 9.40). In cyclic esters (lactones) the $\nu_{C=O}$ frequency increases with ring size decrease: 1750 cm^{-1} for six-membered rings (δ-lactones), 1790–1760 cm^{-1} for five-membered rings (γ-lactones) and near 1840 cm^{-1} for four-membered rings (β-lactones). The $\nu_{C=O}$ band is very strong in the infrared spectrum but of medium strength in the Raman spectrum. It is often split into a doublet in the spectra of α,β-unsaturated lactones owing to Fermi resonance with an overtone of the $\gamma_{=C-H}$ mode. The ester group is also characterized by two bands attributed to the C—O stretching vibration. The higher

Table 4.26—The $\nu_{C=O}$ band positions in the spectra of esters*

Group	Band position, cm^{-1}
HCOOR	1725–1720
RCOOR[a]	1740–1730
C=CCOOR	1730–1715
RCOOC=C	1770–1760
ArCOOR[b]	1720
ArCOOAr	1750
RCOSR	1720–1670
ClCOOR	1780–1770
H_2NCOOR	1695–1690
$ClCH_2COOR$	1750–1740
$Cl_2CHCOOR$	1755–1750
$(COOR)_2$	1760
$CO(OR)_2$	1780–1750
O, =O	1840
O, =O	1790–1770
O, =O	1790–1780 1765–1740
O, =O	1820–1800
O, =O, O	1850–1800

* The band is very strong in the infrared spectrum and medium in the Raman spectrum.
[a] Cf. spectrum in Fig. 9.40.
[b] Cf. spectrum in Fig. 9.41.

Table 4.27—Bands characteristic of the skeletal vibration of esters

Group	Band position,* cm^{-1}			
	ν_{C-O} as	ν_{C-O} sym	$\nu_{R-C=O}$	$\delta_{O=C=O}$
HCOOR	1200–1180 IR	950–850 IR	1385–1375 R	775–750 R
CH_3COOR	1250–1230 IR	1050–1030 IR	850–825 R	645–635 R
RCOOR[a]	1200–1160 IR	1060–1030[b] IR	870–845 R	610–600 R
C=CCOOR	1300–1200 IR		850–800 R	
ArCOOR	1300–1250 IR	1150–1100 IR		

* IR—strong in the infrared spectrum, R—strong in the Raman spectrum.

[a] Cf. spectrum in Fig. 9.40.

[b] In esters of primary or secondary alcohols the band appears in the 1160–1100 cm^{-1} region.

frequency (antisymmetric) falls near 1200 cm^{-1} and is very strong in the infrared spectra of aliphatic esters (generally stronger than the $\nu_{C=O}$ band) (Fig. 9.40). The band is weak in the Raman spectrum. In the spectra of esters of aromatic acids the higher frequency band (near 1300 cm^{-1}) is very strong in the infrared spectrum, whereas the lower frequency band (near 1120 cm^{-1}) is fairly intense in the infrared spectrum and of medium intensity in the Raman spectrum. There are, moreover, other bands which can be distinguished in the Raman spectrum. These bands are attributed to the ν_{C-C} vibration (near 850 cm^{-1}), the $\delta_{O=C-C}$ vibration (650–600 cm^{-1}) and the $\delta_{\underset{\|}{C}-O-C}$ (with =O on the first C) vibration (450–300 cm^{-1}) (Tables 4.26 and 4.27).

Acid anhydrides. Acid anhydrides are characterized by two bands attributed to the antisymmetric and symmetric vibrations of two carbonyl groups. Their frequencies are considerably higher than those of other carbonyl compounds and the interval between the bands is 50–80 cm^{-1}. In non-cyclic saturated anhydrides, one band falls near 1820 cm^{-1}, the other near 1750 cm^{-1} (Fig. 9.42). The relative intensity of those bands is also characteristic. The higher-frequency band is the stronger in both the infrared and Raman spectrum for non-cyclic anhydrides, whereas for cyclic anhydrides (Fig. 9.43) the lower-frequency band is stronger in the infrared spectrum, but in the Raman spectrum both bands have similar intensities. From polarization ratio measurements, the higher-frequency band, which is polarized, may be assigned to the symmetric vibration and the other, depolarized, band to the antisymmetric vibration. Spectra of non-cyclic anhydrides contain, in addition, a strong band at 1100–1000 cm^{-1}, the position of which differs in the infrared spectrum from that in the Raman spectrum (the antisymmetric and symmetric vibrations of the C—O—C group). In the spectra of cyclic anhydrides there are also two quite intense infrared bands, in the 1300–1200 cm^{-1} and 950–900 cm^{-1} regions, respectively. Only the former appears in the Raman spectrum, in which there is an additional strong band, attributed to the anhydride breathing vibration, in the 650–620 cm^{-1} region (Table 4.28).

Acid halides. The $\nu_{C=O}$ band, which is characteristic of acid halides, lies at a particularly high frequency, since the strong electron-withdrawing inductive effect of halogens tends to give rise to an increase in the C=O bond order. Aliphatic acid chlorides, bromides and iodides absorb in the 1810–1795 cm^{-1} region, whereas acid fluorides absorb in the region 1840–1830

Table 4.28—Bands characteristic of acid anhydrides

Group	Type of vibration	Band position, cm^{-1}	Intensity R	Intensity IR
$(RCO)_2O$[a]	$\nu_{C=O}$ sym	1825—1810	s	vs
	$\nu_{C=O}$ as	1755–1745	m	s
$(C{=}C{-}CO)_2O$ }	$\nu_{C=O}$ sym	1780–1770	s	vs
$(ArCO)_2O$ }	$\nu_{C=O}$ as	1725–1715	m	s
Cyclic five-membered:				
saturated	$\nu_{C=O}$ sym	1875–1865	s	s
	$\nu_{C=O}$ as	1800–1780	s	vs
unsaturated[b]	$\nu_{C=O}$ sym	1865–1850	s	s
	$\nu_{C=O}$ as	1780–1770	s	vs
Cyclic six-membered:				
saturated	$\nu_{C=O}$ sym	1820–1800	s	vs
	$\nu_{C=O}$ as	1770–1750	s	vs
unsaturated	$\nu_{C=O}$ sym	1800–1780	s	s
	$\nu_{C=O}$ as	1740–1730	s	vs
$(CH_3CO)_2O$	ν_{C-O} as	1125	w	vs
	ν_{C-O} sym	1000	s	w
$(RCO)_2O$	ν_{C-O} as	1100–1040	w	s
	ν_{C-O} sym	1100–1040	s	w
Cyclic	ν_{C-O} as	1300–1200	s	s
	ν_{C-O} sym	950–900		vs
	ν_{ring}	650–620	vs	
Cyclic saturated	ν_{ring}	1100–1050		vs

[a] Cf. spectrum in Fig. 9.42.
[b] Cf. spectrum in Fig. 9.43.

cm^{-1}. In α,β-unsaturated acid halides the $\nu_{C=O}$ band is about 30 cm^{-1} lower and often appears as a doublet, since the molecule exists in two conformations (antiperiplanar and synperiplanar). This is illustrated by the spectrum in Fig. 9.44. The doublet that appears in the spectra of aromatic acid halides is attributed to Fermi resonance of the $\nu_{C=O}$ mode with a skeletal overtone, which gives rise to an intense band at 890–850 cm^{-1} in the infrared spectrum. In acid chlorides, the ν_{C-Cl} band lies in the 590–565 cm^{-1} region; in bromides the ν_{C-Br} band appears in the 570–530 cm^{-1} region. In the spectra of derivatives of aromatic acids this band appears in the 870–850 cm^{-1} region. The ν_{C-Cl} band is fairly intense in both spectra (Table 4.29).

Table 4.29—Bands characteristic of acid halides

Group	Type of vibration	Band position, cm^{-1}	Intensity R	Intensity IR
RCOF	$\nu_{C=O}$	1840–1830	s	vs
ArCOF	$\nu_{C=C}$	1810–1800	s	vs
RCOCl (Br)	$\nu_{C=O}$	1810–1790	s	vs
C=CCOCl[a]	$\nu_{C=O}$	1780–1750	s	vs
ArCOCl	$\nu_{C=O}$	1785–1765	s	vs
ArCOCl	$2\nu_{C-C}$	1750–1735	m	s
ArCOCl	ν_{C-Cl}	890–850	m	s
RCOCl	ν_{C-Cl}	600–565	m	s
	$\delta_{O=C-Cl}$	440–420	s	m
RCOBr	ν_{C-Br}	570–530	m	s
	$\delta_{O=C-Br}$	360–320	s	m

[a] Cf. spectrum in Fig. 9.44.

Isocyanates. Isocyanates absorb extremely strongly in the infrared, in the 2280–2260 cm^{-1} region. The band is attributed to the antisymmetric mode of the —N=C=O vibration (Fig. 9.45). It is very weak in the Raman spectrum. The symmetric mode of the isocyanate group vibration tends to give rise to a weak and uncharacteristic band.

4.2.4 Halogen compounds

Carbon–halogen stretching bands are poor diagnostic tools since they appear in a region in which there are numerous other bands. In addition, the C—X vibrations are coupled with skeletal vibrations and thus their frequency depends strongly on conformation of the molecule. Nevertheless, if it is known that the molecule of interest contains a halogen, it is generally possible to identify the halogen type from the spectrum, because ν_{C-X} bands are intense in both the Raman (with exception of the ν_{C-F} bands) and infrared spectra. The position of the ν_{C-X} bands also allows identification of the molecular conformation. This is often useful in the conformational analysis of cyclic systems, because the valence frequency for a halogen in the equatorial position is 30–50 cm^{-1} higher than that for a halogen in the axial position. In non-cyclic systems, the ν_{C-X} frequency for conformations in which the halogen atom is antiperiplanar to the carbon atom is 60–100 cm^{-1} higher than that for conformations with the halogen atom synclinal (Table 4.30). In the spectra of halobenzenes, the characteristic

Table 4.30—The ν_{C-X} band positions* in the spectra of halogen compounds†

Type of compound	Band position, cm^{-1} X=Cl	X=Br	X=I
RCH_2X			
sc conformation	660–650[a]	565–560	510–500
ap conformation	730–720	650–640	600–590
R_2CHX			
sc conformation	640–600	540–530[b]	495–485
sc conformation	690–670		
ap conformation	760–740	700–680	585–575
R_3CX			
sc conformation	580–560	520–510	495–485
ap conformation	630–610	590–580	580–570
Cyclopentyl—X			
equatorial position	630–620	520–510	
axial position	600–580		
Cyclohexyl—X			
equatorial position	760–725	710–680	670–650[c]
axial position	715–680	690–650	
$RCHX_2$	sh. 700 as, sh. 650 sym[d]	sh. 600 as, sh. 550 sym	550 as, 465 sym
RCX_2R	sh. 650 as, sh. 560 sym	sh. 585 as, sh. 480 sym	550 as, 450 sym
CHX_3	760 as, 665 sym[e]	655 as, 535 sym	
CHX—CHX			
ap conformation	745(R), 710(IR)	660(R), 590(IR)	
sc conformation	680–650	580–550	

* Bands intense in both Raman and infrared spectra.

† There is an IR band at 1400–1000 cm^{-1} for the C—F group (cf. spectra in Figs 9.40 and 9.44) and there are bands at 1360–1300 and 1060–860 cm^{-1} for the CF_2 group. For the =CF group the Raman band is at 1260–1200 cm^{-1}.

[a] Cf. spectrum in Fig. 9.28.

[b] Cf. spectrum in Fig. 9.47.

[c] Cf. spectrum in Fig. 9.48.

[d] The antisymmetric vibration gives rise to a strong band in the infrared spectrum, and the symmetric to a strong band in the Raman spectrum.

[e] Cf. spectrum in Fig. 9.46.

bands do not correspond to those for aliphatic compounds. Halobenzenes are characterized by an intense band in the infrared and Raman spectra, which is attributed to the skeletal vibration accompanied by the C—X stretching vibration. This band appears in the 1270–1160 cm^{-1} region in the spectra of fluorine derivatives and the 1090–1030 cm^{-1} region in the spectra of other halogen derivatives (Table 4.31 and Figs. 9.23, 9.25 and 9.49). Its position also depends on the substitution of the benzene ring.

Table 4.31—Positions of the characteristic band in the spectra of halobenzenes

Group	Band position, cm^{-1}
Ar—F[a]	1270–1160
p-Ar—Cl	1090
m-Ar—Cl	1075
o-Ar—Cl[b]	1045
p-Ar—Br	1070
m-Ar—Br	1070
o-Ar—Br	1035
p-Ar—I[c]	1060

[a] Cf. spectrum in Fig. 9.43.
[b] Cf. spectrum in Fig. 9.25.
[c] Cf. spectrum in Figs. 9.23, 9.49.

Halogens also influence the frequencies of other vibrations. For instance, halogens cause the $\nu_{C—H}$ frequency to increase (sometimes to above 3000 cm^{-1}) and the $\delta_{C—H}$ and $\omega_{C—H}$ frequencies to decrease (cf. Table 4.4, p. 96). Bromine and iodine cause the $\nu_{C=C}$ frequency to decrease, whereas fluorine raises this frequency (cf. Table 4.6, p. 100). Halogens also shift the $\nu_{C=O}$ vibration to higher frequencies.

4.2.4.1 Fluorine derivatives of hydrocarbons

The C—F stretching band is very strong in the infrared spectrum but very weak in the Raman spectrum. It is located in the 1400–1000 cm^{-1} region (Fig. 9.40).

4.2.4.2 Chlorine derivatives of hydrocarbons

The $\nu_{C—Cl}$ band of aliphatic chlorides appears in the 800–600 cm^{-1} region; it is strong in both the Raman and infrared spectra. Primary alkyl chlorides (Fig. 9.28) absorb near 725 cm^{-1} (antiperiplanar conformation) and near 660 cm^{-1} (synclinal conformation). Secondary alkyl chlorides absorb near 750 cm^{-1} (antiperiplanar conformation) and at 680 and 620 cm^{-1} (synclinal conformation). Tertiary alkyl chlorides absorb near 620 cm^{-1} (antiperiplanar conformation) and 570 cm^{-1} (synclinal conformation). Cyclohexane derivatives exhibit a band close to 750 cm^{-1} (equatorial chlorine atom) and near 700 cm^{-1} (axial chlorine atom). In compounds in which the chlorine atom is attached to the C=C bond, the C—Cl bond is stronger than normal (because of the mesomeric effect), and thus the $\nu_{C—Cl}$ band falls near 850 cm^{-1} (Fig. 9.44). Compounds with several geminal chlorine atoms are characterized by a strong antisymmetric infrared band in the 760–650 cm^{-1} region and a strong symmetric Raman band in the 700–560 cm^{-1} region (Fig. 9.46).

4.2.4.3 Bromine derivatives of hydrocarbons

The $\nu_{C—Br}$ band of aliphatic bromides is located in the 670–515 cm^{-1} region. Spectra of primary alkyl bromides contain a band near 645 cm^{-1} (antiperiplanar conformation) and a band near 560 cm^{-1} (synclinal conformation). Secondary alkyl bromides absorb near 660 cm^{-1} (antiperiplanar conformation) and close to 540 and 580 cm^{-1} (synclinal conformation). Tertiary alkyl bromides absorb near 585 cm^{-1} (antiperiplanar conformation) and close to 515 cm^{-1} (synclinal conformation). Cyclohexane derivatives with an equatorial bromine atom exhibit a band near 690 cm^{-1}, whereas those with an axial bromine atom absorb close to 660 cm^{-1}.

4.2.4.4 Iodine derivatives of hydrocarbons

The C—I stretching vibration is assigned to a band in the 610–485 cm^{-1} region. Spectra of primary alkyl iodides contain a band near 595 cm^{-1} (antiperiplanar conformation) and a band near 505 cm^{-1} (synclinal conformation). Spectra of secondary and tertiary alkyl iodides have bands ($\nu_{C—I}$) located in similar regions: at 575 cm^{-1} (antiperiplanar conformation) and 490 cm^{-1} (synclinal conformation). In cyclohexane derivatives, for example in iodosterols, the $\nu_{C—I}$ band falls near 670 cm^{-1} (Fig. 9.48).

4.2.5 Sulphur compounds

4.2.5.1 Thiols, sulphides and disulphides

The Raman spectra are particularly useful for identifying sulphur compounds, since they have intense characteristic bands, which are absent in the infrared spectrum. Thiols (mercaptans) are characterized by a strong band in the 2600–2550 cm^{-1} region, which is attributed to the $\nu_{S—H}$ stretching vibration (Fig. 9.50). The band is very weak in the infrared spectrum. Another characteristic band of thiols is due to the ν_{C-S} stretching vibration. It appears in the 700–600 cm^{-1} region and is strong in the Raman spectrum (Fig. 9.50). This band is also characteristic of sulphides and disulphides (Figs. 9.36 and 9.51). The Raman spectra of disulphides also contain a very intense band attributed to the S—S stretching vibration in the 520–450 cm^{-1} region (Figs. 9.36 and 9.51).

4.2.5.2 Sulphoxides

The S=O stretching frequency, which is characteristic of sulphoxides, lies close to 1050 cm^{-1}. The $\nu_{S=O}$ band is very strong in the infrared spectrum but of medium strength in the Raman spectrum. On the other hand, the ν_{C-S} band in the 700–600 cm^{-1} region is strong in the Raman spectrum (cf. the dimethyl sulphoxide spectrum in Fig. 9.52). The $\nu_{S=O}$ frequency increases with increasing electronegativity of the atoms bonded to sulphur. In alkyl sulphites, for example, the band falls in the 1225–1200 cm^{-1} region.

4.2.5.3 Sulphonyl compounds

The SO_2 group is characterized by two bands attributed to the symmetric and antisymmetric vibrations of the coupled S=O oscillators. The ν_{SO_2} as band lies in the 1400–1300 cm^{-1} region and the ν_{SO_2} sym band in the 1200–1100 cm^{-1} region (Fig. 9.53). The S=O stretching frequency increases with increasing electronegativity of the substituents attached to the sulphonyl group. Both bands are very strong in the infrared spectrum, whereas in the Raman spectrum only the symmetric band is strong (Table 4.32).

4.2.6 Phosphorus compounds

Organophosphorus compounds and organic esters of phosphoric and thiophosphoric acids are useful as specific reagents in organic synthesis and as biologically active compounds. The band positions for the principal groups in phosphorus compounds are listed in Table 4.33. The P—H stretching vibration appears as a band in the 2440–2275 cm^{-1} region. It is intense in the Raman spectrum but of medium intensity in the infrared spectrum. The band assigned to the valence vibration of the phosphoryl group (P=O) falls in the 1300–1140 cm^{-1} region. It is very strong in the infrared spectrum and of medium strength in the Raman spectrum (Fig. 9.54). On the other hand, the thiophosphoryl group gives rise to a strong band only in the Raman spectrum. Nevertheless, the $\nu_{P=S}$ vibration is

Table 4.32—Bands characteristic of sulphur compounds

Group	Type of vibration	Band position, cm^{-1}	Intensity R	Intensity IR
RSH,[a] ArSH	ν_{S-H}	2600–2550	s	vw
	ν_{C-S}	735–590	vs	w
RSR	ν_{C-S} sym	700–585	vs	w
	ν_{C-S} as	750–700	s	s
RSSR[b]	ν_{S-S}	540–510	vs	
	ν_{C-S}	715–620	s	w
$R_2S{=}O$[c]	$\nu_{S=O}$	1060–1040	m	vs
	ν_{C-S}	700–600	s	m
$(RO)_2S{=}O$	$\nu_{S=O}$	1210–1200	s	vs
R_2SO_2	ν_{SO_2} as	1350–1310	w	vs
	ν_{SO_2} sym	1180–1140	s	vs
	ν_{C-S}	700–650	s	
	δ_{SO_2}	580–500	s	w
$(RO)_2SO_2$	ν_{SO_2} as	1420–1380	vw	vs
	ν_{SO_2} sym	1210–1180	s	vs
RSO_2NH_2	ν_{SO_2} as	1370–1300	vw	vs
	ν_{SO_2} sym	1180–1140	s	vs

[a] Cf. spectrum in Fig. 9.50.
[b] Cf. spectrum in Figs. 9.36, 9.51.
[c] Cf. spectrum in Fig. 9.52.
[d] Cf. spectrum in Fig. 9.53.

Table 4.33—Positions of bands corresponding to the stretching vibrations in organophosphorus compounds

Group	Band position, cm^{-1}	Intensity R	Intensity IR
P—H	2440–2275	s	m
P—C	770–650	s	m
P—Ar	1130–1090	m	s
P—Cl	580–440	s	s
$P{-}CH_3$[a]	1310–1280	m	s
P=S	850–750	s	m
P—O—R	1050–970 as	w	vs
	830–740 sym	s	s
P—O—Ar	1260–1160 as	w	vs
	995–855 sym	s	s
P—N—C	1190–930 as	w	vs
	770–680 sym	s	m
$R_3P{=}O$	1190–1150	m	vs
$(RO)_3P{=}O$	1300–1250	m	vs

[a] The δ_{CH_3} sym vibration.

more strongly coupled with other vibrations (e.g. with the ν_{P-C} or ν_{P-O} mode) than is the $\nu_{P=O}$ vibration. The oscillators of the P—O—C grouping in esters as well as the oscillators of the P—N—C grouping in amides are strongly coupled, so these groupings undergo two vibrations, an antisymmetric and a symmetric. The interval between the vibration frequencies is about 250–200 cm^{-1}. The symmetric frequency is lower and the band strong in the Raman spectrum, whereas the higher-frequency band (antisymmetric) is very strong in the infrared spectrum. The ν_{P-C} band appears in the 770–650 cm^{-1} region and is strong in the Raman spectrum. However, if several hydrocarbon groups are bonded to the same phosphorus atom, the ν_{P-C} sym band and the ν_{P-C} as band will appear at different frequencies, as is also the case for the P—O—C or the P—N—C group. The P—Cl group is characterized by a strong band in both the Raman and infrared spectra.

4.2.7 Silicon compounds

Silicone oils, lubricants, paints and gums have found a very wide application in various areas of industry. Thus methods of spectral identification of particular groups present in silicone compounds are of general interest. The siloxane group is the fragment that occurs most frequently in silicones. The Si—O—Si group is characterized, as is the ether group, by an antisymmetric and a symmetric vibration. The former leads to a strong infrared band (1225–1010 cm^{-1}), the latter is strong in the Raman spectrum (540–460 cm^{-1}). It is perfectly illustrated by the hexamethyldisiloxane spectrum (Fig. 9.55). Mixed ethers (Si—O—C) also have an antisymmetric (1190–930 cm^{-1}) and a symmetric vibration (850–800 cm^{-1}). The methyl group bonded to Si, which often occurs in silicones, gives rise to a characteristic infrared band assigned to the δ_{CH_3} sym symmetric mode in the 1300–1250 cm^{-1} region. The Si—H stretching vibration in silanes leads to a strong Raman band and a medium strength infrared band in the 2250–2100 cm^{-1} region. Silanols, similarly to alcohols, absorb in the 3700–3200 cm^{-1} region (ν_{Si-OH}). This band is strong in the infrared spectrum. The ν_{Si-O} band in the 910–835 cm^{-1} region is also strong in the infrared spectrum. The valence vibrations ν_{Si-C} are usually assigned to strong infrared bands at 850–760 cm^{-1} and strong Raman bands in the 760–560 cm^{-1} region. The great divergence in the band positions is caused by the fact that in silicones two alkyl groups (usually methyl groups) are often attached to a single silicon atom. This results in an antisymmetric

ν_{Si-C} as (infrared band), and a symmetric mode ν_{Si-C} sym (Raman band). The Si—Cl stretching vibration in alkyl chlorosilanes, used as intermediates in the synthesis of silicones, gives rise to strong bands in both Raman and infrared spectra. The band lies in the 620–450 cm^{-1} region. If two or more chlorine atoms are attached to the same silicon atom, the ν_{Si-Cl} as band is observed in the upper range of the given spectral region, whereas the ν_{Si-Cl} sym band is located in its lower range. The antisymmetric vibration gives a strong infrared band, whereas the symmetric leads to a strong band in the Raman spectrum (Table 4.34).

Table 4.34—Bands characteristic of silicon compounds

Group	Type of vibration	Band position, cm^{-1}	Intensity R	Intensity IR
Si—OH	ν_{O-H}	3700–3200	w	s
Si—H	ν_{Si-H}	2250–2100	s	ms
	δ_{Si-H}	950–910	m	s
Si—C	ν_{Si-C} as	850–760	w	s
	ν_{Si-C} sym	760–560	vs	w
Si—CH_3	δ_{CH_3} sym	1280–1255	m	s
Si—Si	ν_{Si-Si}	430–400	vs	
Si—O—Si[a]	ν_{Si-O} as	1125–1010	w	vs
	ν_{Si-O} sym	540–460	vs	w
Si-O-R	ν as	1190–1100	m	vs
	ν sym	850–800	s	s
Si—O—Ar	ν_{Si-O}	970–920	m	s
Si—Cl	ν_{Si-Cl}	620–450	s	s

[a] Cf. band in Fig. 9.55.

4.2.8 Organometallic compounds

Organometallic compounds can rarely be identified by using the methods of vibrational spectroscopy, for several reasons. If a light metal atom, for example, beryllium, aluminium or magnesium, forms a bond with a carbon atom, then the carbon–metal valence vibration is strongly coupled with other skeletal vibrations and is therefore of little value for purposes of identification. If a heavy metal atom, e.g. mercury, lead, zinc or tin is linked to a carbon atom, then the carbon–metal stretching modes are responsible for bands in the crowded 600–450 cm^{-1} region. Although these modes tend to give rise to strong bands in the Raman and infrared spectra, it is not always possible to identify the metal present in the molecule.

An additional difficulty is the identification of the bridge dimer form in which many organometallic compounds exist, particularly those containing aluminium or beryllium or chloro-organometallic compounds. In this case, the vibrational frequencies of the carbon–metal or chlorine–metal bonds may vary, depending on whether the hydrocarbon group or the chlorine atom is located outside the bridge, or whether it forms the bridge itself (Table 4.35). Therefore neither the Raman spectrum nor the infrared

Table 4.35—Position of bands* corresponding to carbon–metal valence vibrations

Group	Band position, cm^{-1}
Be—C	1100–450
Al—C	700–450[a]
Mg—C	500–300[a]
$R_3Sn—SnR_3$[b]	210–190
RHg—HgR[b]	180–160
Pb—C	460–440
Sn–C	520–480
Hg—C	550–500
Zn—C	500
Ge—C	600—550

* Bands intense in both infrared and Raman spectra.
[a] Also the valence vibration of bridge bonds.
[b] The ν_{M-M} bands are strong only in the Raman spectrum.

spectrum is ever used for the identification of an organometallic compound. These spectroscopic methods are principally useful for examining spectral characteristics of organometallic compounds, determining their geometric structure and defining the character of the carbon–metal bond. Raman spectra are better suited for the identification of vibrational frequencies in the lower range than are infrared spectra and are also more easily obtained for compounds sensitive to water and oxygen.

REFERENCES

[1] F. R. Dollish, W. G. Fateley and F. F. Bentley, *Characteristic Raman Frequencies of Organic Compounds*, Wiley-Interscience, New York, 1974.
[2] M. C. Tobin, *Laser Raman Spectroscopy*, Wiley-Interscience, New York, 1971.
[3] T. R. Gilson and P. J. Hendra, *Laser Raman Spectroscopy*, Wiley-Interscience, London, 1970.
[4] G. Brandmüller and G. Moser, *Einführung in die Raman Spektroskopie*, Steinkopff Verlag, Darmstadt, 1962.

[5] N. B. Colthup, L. H. Doly and S. E. Wiberley, *Introduction to Infrared and Raman Spectroscopy*, Academic Press, New York, 1964.
[6] C. N. R. Rao, *Chemical Applications of Infrared Spectroscopy*, Academic Press, New York, 1963.
[7] A. Finch, P. N. Gates, K. Radcliffe, F. N. Dickson and F. F. Bentley, *Chemical Applications of Far Infrared Spectroscopy*, Academic Press, London, 1970.
[8] J. Woliński and J. Terpiński, *Organic Qualitative Analysis* (in Polish), PWN, Warszawa, 1973.
[9] B. Schräder and W. Meier, *Raman IR-Atlas*, Verlag Chemie, Weinheim, 1974.

5

Identification of inorganic compounds

J. Terpiński

The relation between structure and band position and intensity for inorganic compounds (arising from particular vibrations of ions or neutral molecules), is the same as that for organic compounds (see Chapter 4). As is the case for organic compounds, the stretching frequencies of double bonds are higher than those of single bonds and give rise to stronger Raman bands. In addition, the symmetric vibrations tend to give rise to stronger bands in the Raman spectrum, whereas the antisymmetric vibrations give stronger infrared bands. Nevertheless, the analysis of vibrational spectra (both Raman and infrared) of inorganic compounds presents some specific problems which do not occur in the investigation of the spectra of organic compounds. One of these problems is caused by a fundamental difference between organic and inorganic compounds. Organic compounds contain many similar structural elements and only a few functional groups, which are easily identifiable by spectroscopic methods owing to their characteristic bands. On the other hand, inorganic compounds are formed from a much wider range of elements, with widely differing masses and bond strengths. As the vibrational frequency depends on the masses of the vibrating atoms and the force constant of the bond between them, the structural elements of an inorganic compound can rarely be identified without some prior knowledge of the chemical structure of the compound. As a result, the value of a spectroscopic examination in the identification of inorganic compounds is strictly limited. A further problem is posed by the fact that the bonding in inorganic compounds differs from that in organic compounds. In organic compounds, the bonding is substantially covalent, with strong intramolecular interactions, which in solids are almost independent of the intermolecular interactions. Thus, the concept of localized vibrations can be used in the identification of particular structural elements. In contrast, in inorganic compounds the bonding is mainly ionic in nature, so that in the solid state the entire crystal, made up of cations and anions, may be considered as a 'supermolecule' without localized vibrations. Nevertheless, there are a large number of simple salts formed with anions (e.g. NO_3^-, SO_4^{2-}) which consist of only a few atoms and have well-defined steric structures with a high degree of symmetry. In such ions the interactions between the atoms are much stronger than those between the ion and other ions in the crystal. Thus the vibrations of such ions are characteristic of their structure and can be used for their identification. In order to eliminate specific interactions between ions in the crystal lattice, the spectroscopic examination should be done on aqueous solution. Here, Raman spectroscopy has several advantages over infrared spectroscopy: optical cells may be made of glass, light-scattering by water is low, whereas the infrared absorbance is considerable [cf. the spectra in Fig. 9.56 (H_2O) and Fig. 9.57 (D_2O)] and the sharp bands observed in Raman spectra make interpretation easier. For these reasons, Raman spectroscopy is frequently used for detecting and identifying complex inorganic ions. Monatomic ions, such as metal cations or halogen anions, cannot of course be identified by Raman spectra. However, if they can be converted into complex ions, by adding a suitable complexing agent, the resulting Raman spectra can then be used for identifying the monatomic ions. As an example, if a solution containing Al^{3+} cations is saturated with potassium chloride, complex $AlCl_4^-$ ions are formed, resulting in the appearance of intense bands in the Raman spectrum due to the aluminium–chlorine stretching vibration ($\nu_{Al—Cl}$).

Moreover, if the structure of the ion is simple and contains only a few atoms then group theory and symmetry rules may be applied to the analysis of the vibrational spectrum to a greater extent than is the case with organic compounds. If the symmetry group of an ion is known, it is easy to compute the number of vibrations and predict which will give rise to Raman bands and which to infrared bands. It can also be found whether a particular band is polarized or unpolarized. From the analysis

Table 5.1—Positions of the characteristic bands* of the important tri- tetra-,

Ion (point group)	ν_1 band, cm^{-1} (type of vibration, symmetry type)				ν_2 band, cm^{-1} (type of vibration, symmetry type)	
	in H_2O	in crystals	intensity R	intensity IR	in H_2O	in crystals
$D_{\infty h}$		νsym (Σ^+)				π
N_3^-	1350	1380–1320	vs		635	660–630
NO_2^+	1400		vs		570	
FHF^-		610–585	vs		1205	1260–1200
$C_{\infty v}$						
OCN^-	1292, 1205	1215–1200	s	w	615	640–600
SCN^- [a]	745		s	w	470	490–420
C_{2v}		νsym(A_1)				$\delta(A_2)$
NO_2^-	1330	1380–1320	s	s	815	835–825
D_{3h}		νsym(A_1')				$\pi(A_2'')$
BO_3^{3-}	910 p	1000–925	vs		700	760–680
NO_3^- [b]	1050 p	1060–1020	vs		825	835–780
CO_3^{2-} [c]	1060 p	1090–1050	vs		880	890–850
C_{3v}		νsym(A_1)				$\delta(A_1')$
SO_3^{2-}	965 p	990–950	vs	s	615 p	650–620
ClO_3^-	930 p	935–915	s		625 p	625–600
BrO_3^-	800 p	805–770	s		440 p	445–420
IO_3^-	780 p	780–695	s		390 p	415–400
OH_3^+	3380–3780		m	vs	1180–1150	
T_d		νsym(A_1)				$\delta(E)$
NH_4^+ [d]	3040 p		s		1680	
ND_4^+	2215 p		s		1215	
PO_4^{3-}	935 p	975–960	vs		565	600–540
SO_4^{2-} [e]	980 p	1010–970	vs		615	680–610
ClO_4^-	930 p	940–930	vs		625	
IO_4^-	790 p		vs		325	
MnO_4^-	840 p	860–840	vs		430	
CrO_4^-	850 p	860–840	vs		370	
AsO_4^{3-}	810 p		vs		340	
AlH_4^-	1740 p	1640	s	s	800	900–800
BF_4^-	770 p		vs	s	525	

* The letter p (after the wavenumber) denotes that the band is polarized.

[a] Cf. spectrum in Fig. 9.58.

[b] Cf. spectrum in Fig. 9.59.

[c] Cf. spectrum in Fig. 9.60.

[d] Cf. spectra in Figs. 9.58, 9.62.

[e] Cf. spectrum in Fig. 9.61.

and penta-atomic inorganic ions in aqueous solution and in crystalline salts

		ν_3 band, cm^{-1} (type of vibration, symmetry type)				ν_4 band, cm^{-1} (type of vibration, symmetry type)			
intensity		in H_2O	in crystals	intensity		in H_2O	in crystals	intensity	
R	IR			R	IR			R	IR
			νas(Σ^+)						
		2070	2190–2010		vs				
		2360			vs				
m	m	1535	1700–1400		vs				
	m	2190 p	2220–2130	s	s				
w	w	2065 p	2160–2040	vs	s				
			νas(B_1)						
m	s	1230	1260–1230	m	vs				
			νas(E')				$\delta(E')$		
	s		1460–1240		vs		680–590	m	m
w	s	1400		m	vs	720	740–710	m	w
	m	1415	1495–1380	w	vs	680	740–610	w	m
			νas (E)				δ (E)		
w	m	950	970–890	m	s	470	520–470	s	m
		980	990–965	m	s	480	500–460		
		830	830–800	m	s	350			
w	w	825	830–760	m	s	350			
m	m	3380–3270		m	vs	1700–1600		m	s
			νas (F_2)				δ (F_2)		
		3145			s	1400	1430–1390		s
		2350			s	1065			
w	m	1080	1100–1080		s	420	500–400		
w	m	1100	1140–1080	w	vs	450		w	
		1130	1140–1080			460			
		850				255			
		920	940–880			355			
		885	915–870			350			
		810			vs	400			
w	m	1740	1785		vs	745	800–700	s	s
w	s	985		s	vs	355		w	s

Table 5.2—Position of bands in spectra

Ion	Type of vibration and band position, cm^{-1}	Intensity R	Intensity IR	Type of vibration and band position, cm^{-1}	Intensity R	Intensity IR	Type of vibration and band position, cm^{-1}
$S_2O_3^{2-}$	$\nu_{S—S}$			ν_{SO_3} sym			ν_{SO_3} as
in H_2O	670	s	s	995	vs	s	1125
in crystals	690–640	s	s	1010–990			1150–1080
HSO_3^-	$\nu_{S—H}$			ν_{SO_3} sym			ν_{SO_2} as
in H_2O	2615	s	w	1040	vs	s	1195
$HO—SO_3^-$ [a]	$\nu_{S—O}$			ν_{SO_3} sym			ν_{SO_3} as
in H_2O	885		s	1050	vs	vs	1200
in crystals	900–850		s	1080–1020	vs	vs	1300–1170
HPO_3^{2-}	$\nu_{P—H}$			ν_{PO_3} sym			ν_{PO_3} as
in H_2O	2315	s	m	980	s	s	1085
				990–920	s	s	1095–970
$HO—PO_3^{2-}$ [a,b]	$\nu_{P—O}$			ν_{PO_3} sym			ν_{PO_3} as
in H_2O	890		s	970	vs	s	1085
in crystals	860		s	990	vs	s	1150–1040
$(HO)_2PO_2^-$ [a]	$\nu_{P—O}$ sym			ν_{PO_2} sym			ν_{PO_2} as
in H_2O	875	s		1070	s	vs	1150
$HO—CO_2^-$ [a,c]	$\nu_{C—O}$			ν_{CO_2} sym			ν_{CO_2} as
d	960		m	1340	vs	vs	1696
in crystals[e]	1050–970		m	1450–1350	s	vs	1680–1615
$O_3POPO_3^{4-}$	ν_{PO_3} sym			ν_{PO_3} as			ν_{OPO} sym
in crystals	1025–985	s	s	1150–1120	s	vs	730
$O_3CrOCrO_3^{2-}$	ν_{CrO_3} sym			ν_{CrO_3} as			ν_{OCrO} sym
in crystals	910–840	s	s	965–925	s	vs	570–555

[a] These ions have broad infrared bands with a series of maxima in the 3400–2000 cm^{-1} region. The
[b] Cf. spectrum in Fig. 9.62.
[c] Cf. spectrum in Fig. 9.63.
[d] Data for a free ion in a low temperature matrix.
[e] It is supposed that in crystals the hydrogen carbonate ions form cyclic dimers, similar to organic

of ions of more complex structure

Intensity R	Intensity IR	Type of vibration and band position, cm^{-1}	Intensity R	Intensity IR	Type of vibration and band position, cm^{-1}	Intensity R	Intensity IR	Type of vibration and band position, cm^{-1}	Intensity R	Intensity IR
	s	δ_{SO_3} sym			δ_{SO_3} as			δ_{SSO}		
	vs	445			540			335	m	
					570–540					
		δ_{SO_3} sym			δ_{SO_3} as			δ_{HSO}		
m	vs	510	s	s	625	m	s	1135	m	s
		δ_{SO_3} sym			δ_{SO_3} as			δ_{OSO}		
	vs	595	s		595		s	430	s	
	vs	620–565		s	620–656		vs	480–450	s	
		δ_{PO_3} sym			δ_{PO_3} as			δ_{HPO}		
	vs	565		m	565		m	1025		
	vs									
		δ_{PO_3} sym			δ_{PO_3} as					
	vs	530			530					
	vs	535			535					
		δ_{PO}			τ			δ_{OH}		s
m	vs	935		s	1030			1300		
		δ_{CO}			δ_{CO}			π		
m	vs	710		m	580		w	835	m	
m	vs	680–635		m						
		ν_{OPO} as			δ_{PO_3} sym			δ_{PO_3} as		
s	w	915		s	520–415	s	m	560	w	s
		ν_{OCrO} as			δ_{CrO_3} sym			δ_{CrO_3} as		
s	w	795–765	w	s	365	s	m		w	s

bands are associated with the ν_{O-H} vibrations and overtones of the hydroxyl group.

acids.

of both the Raman and infrared spectra of an ion of known constitution, a guess can be made at its symmetry and hence at its steric structure, without an arduous X-ray diffraction study. A further development of this method is the identification of crystal symmetry from the differences between the spectra for the free ion and the crystal. The simplest approach consists in first considering a set of ions of known symmetry which form the unit cell in the crystal, and then assuming that the cell belongs to one of the recognized symmetry classes. Thus, the number of vibrations for the given set of ions can be computed and the selection rules which govern the activity of the vibrations can be determined. Finally, a check can be made to see whether the characteristics of the recorded spectrum are in agreement with the theoretical assumptions. If they are not, a fresh start is made with calculations for the next possible class of symmetry.

The positions and intensities of the characteristic bands for the most important inorganic ions are listed in Tables 5.1 and 5.2. Additional information is given on the type of symmetry of the vibrations listed and on the regions in which the characteristic ion bands appear in the spectra of crystals. The data given in the tables are illustrated in Figs. 9.58–9.63.

Laser Raman spectroscopy has been widely used in physics in the investigation of the properties of the solid state. There are two main reasons for this. First, it is possible to observe very low frequencies (down to a few tens of cm^{-1}) of vibration in these solids. Secondly, the development of the theory allows prediction of the number and types of low-frequency vibrations allowed in the crystal lattice.

Raman spectroscopy is useful not only for the analysis of substances of ionic character but also for the identification and determination of structures of simple inorganic molecules with covalent bonding. Some of these, for example the centrosymmetric, diatomic molecules of nitrogen, oxygen, chlorine and bromine, have bands observable only in the Raman spectrum, since they do not absorb infrared radiation. The band positions in the spectra of some compounds of this type are listed in Table 5.3 and illustrated by spectra in Figs. 9.64–9.66.

Raman spectroscopy is also useful for the investigation of reactions taking place between compounds present in a sample. It is possible to observe the formation of complexes (see Section 7.2) and new ions, or to determine degrees of dissociation. As an example, the spectrum of pure (100%) nitric acid corresponds to the $HONO_2$ structure, whereas the spectra of diluted aqueous solutions of the acid have bands corresponding to the NO_3^- ion. The degree of dissociation can be determined from the

Table 5.3—Characteristic bands in spectra of common inorganic compounds of non-ionic structure

Compound	Type of vibration	Band position, cm^{-1}	Intensity R	Intensity IR	Type of vibration	Band position, cm^{-1}	Intensity R	Intensity IR	Type of vibration	Band position, cm^{-1}	Intensity R	Intensity IR
H_2O[a]	$\nu_{O—H}$ sym	3450, br	s	vs	$\nu_{O—H}$ as	3450, br	s	vs	δ_{OH_2}	1640	m	vs
D_2O[b]	$\nu_{O—D}$ sym	2550	s	vs	$\nu_{O—D}$ as	2550	s	vs	δ_{OD_2}	1220	m	vs
H_2O_2	$\nu_{O—H}$	3400, br	s	vs	$\nu_{O—O}$	880	vs		$\delta_{O—H}$	1400, 1350		
H_2SO_4[c]	$\nu_{O—H}$	3000, br	w	vs	ν_{SO_2} sym	1140	vs	vs	ν_{SO_2} as	1370	w	vs
	$\nu_{S—O}$ sym	910	vs	s	$\nu_{S—O}$ as	970	w	vs	δ_{SO_2}	560	s	s
	$\delta_{S—O}$	390	s	w								
HNO_3	$\nu_{O—H}$	3400, br	w	s	ν_{NO_2} sym	1300	vs	s	ν_{NO_2} as	1675	m	vs
	$\nu_{N—O}$	925	vs	s	δsym	680	s		δas	610		s
$HClO_4$	$\nu_{O—H}$	3000, br	w	s	ν_{ClO_3} sym	1035	vs	s	ν_{ClO_3} as	1230	w	vs
	$\nu_{Cl—O}$	740	s	s	δ_{ClO_3}	575	s	m				
$SOCl_2$	$\nu_{S—O}$	1230	s	vs	ν_{SCl_2} sym	490	vs	s	ν_{SCl_2} as	440		vs
	δ_{OSCl}	394	m	m	δ_{SCl_2}	195	s					
SO_2Cl_2	ν_{SO_2} sym	1205	vs	s	ν_{SO_2}as	1435	w	vs	ν_{SCl_2} sym	405	vs	s
	ν_{SCl_2} as	580	m	vs	δ_{SO_2}	580			δ_{SCl_2}	210		
PCl_3	$\nu_{P—Cl}$ sym	505	vs	s	$\nu_{P—Cl}$ as	490	w	vs	δsym	200	s	s
	δas	250	m	s								
PBr_3	$\nu_{P—Br}$ sym	380	vs	s	$\nu_{P—Br}$ as	400	w	vs	δsym	115	s	s
	δas	160	m	s								
$POCl_3$	$\nu_{P=O}$	1290	m	s	$\nu_{P—Cl}$ sym	485	vs	s	$\nu_{P—Cl}$ as	580	w	vs
Al_2Cl_6[d]	νsym(t)	505	s	w	νsym(b)	340	s	w	δ_{AlCl_2}	215	s	m
	νas(t)	625	w	s	νas(b)	420	w	s				

[a] Cf. spectrum in Fig. 9.56.
[b] Cf. spectrum in Fig. 9.57.
[c] Cf. spectrum in Fig. 9.65.
[d] (t)—vibration of chlorine atoms in terminal position, (b)—vibration of chlorine atoms in bridge position (cf. p. 140).

ratio of the intensities of bands in the spectra of an acid in solutions of different concentrations (Chapter 6). Moreover, Raman spectroscopy makes it possible to observe the formation of nitronium ion in the course of the reaction of nitric acid with any other strong acid, e.g. perchloric or sulphuric acid.

REFERENCES

[1] S. D. Ross, *Inorganic Infrared and Raman Spectra*, McGraw-Hill, New York, 1972.
[2] G. Brandmüller and G. Moser, *Einführung in die Raman Spektroskopie*, Steinkopff Verlag, Darmstadt, 1962.
[3] N. B. Colthup, L. H. Doly and S. E. Wiberley, *Introduction to Infrared and Raman Spectroscopy*, Academic Press, New York, 1964.

6

Quantitative analysis

H. Barańska

6.1 INTRODUCTION [1–4]

The principal field of application of laser Raman spectroscopy has been in the determination of structure and identification of samples. As was the case with other spectroscopic methods, however, when the technique had developed sufficiently, the demand arose for the spectra to provide quantitative information. This information could be used not only in chemical analysis, but also in the determination of physicochemical constants such as the enthalpies and equilibrium constants of chemical reactions. Because of the comparative ease with which the vibration–rotation bands of molecules can be recorded, Raman spectroscopy may be used for solving such quantitative problems, which often cannot be solved by other more common analytical methods.

The quantitative methods used in Raman spectrometry are, in general, similar to those used in all other procedures of instrumental analysis [5–8]. These procedures involve comparison of spectra for the unknown sample and a standard, i.e. a pure substance, or a mixture of known composition.

From the physical nature of the scattering process it follows that the measured intensity I of the scattered radiation is directly proportional to the number of scattering centres present in the volume illuminated by the laser beam.

From the Placzek theory this intensity is given by the equation:

$$I = KI_0N\frac{(\nu_0-\nu)^4}{\nu(1-e^{-h\nu/kT})}\sum_{ij}\left(\frac{\partial\alpha_{ij}}{\partial Q}\right)^2 \tag{6.1}$$

where I is the intensity of the Stokes Raman band when the Raman radiation is scattered in a gas sample containing N molecules and the observation is made at an angle of 90° to the excitation beam, K is a coefficient with a value which depends only on the units used for the other quantities in the equation, I_0 is the intensity and ν_0 the frequency of the exciting beam, ν is the frequency of the normal vibration of the molecule under consideration, k is the Boltzmann constant, T is the absolute temperature, $\sum_{ij}(\partial\alpha_{ij}/\partial Q)^2$ is the sum of the squares of the derivatives of polarizability with respect to the normal coordinate of vibration Q, over all components of the polarizability tensor.

If all measuring conditions, i.e. I_0, ν_0, T, illuminated volume, detector and recorder sensitivity, are kept constant, then Eq. (6.1) reduces for any given band to:

$$I = a_0c \tag{6.2}$$

where c is the volume concentration, and a_0 is the proportionality coefficient. This coefficient changes on passing from the gas to the condensed phase, in which interactions between the molecules become important. Some authors [9–11] distinguish three separate effects related to molecular interactions in the liquid phase, and these may be expressed as three multiplying factors (G, F, L) which modify the coefficient a_0.

Factor G refers to the *optical effect* which is dependent on the refractive index of the liquid. Refraction of the scattered light in the liquid produces a change of the radiation energy reaching the detector. If a liquid is illuminated by a laser beam and the scattered radiation is observed at right angles to the incident beam we can assume that $G = 1/n^2$, where n is the refractive index of the liquid sample. Factor G is called the *Woodward and George correction factor* [12].

Factor F is due to the *internal field effect in the sample* as a whole (after Onsager [13]). Its value is also related to the refractive index of the liquid, since both the internal field and refractive index result from universal interactions of the molecule with its environment.

Factor L is due to the *effect of specific intermolecular interactions* leading to the formation of complexes, i.e. the creation of bonds between the molecules of the sample.

The *scattering index* for the liquid phase, a_1, is thus given by the equation:

$$a_1 = a_0 GFL \tag{6.3}$$

The experimentally measured intensity of the Raman-scattered radiation I, can be expressed by the equation:

$$I = ac \tag{6.4}$$

where a is a constant characteristic of the given band of a molecule in a specified medium under specified measuring conditions (instrument, sample illumination, detector).

Equation (6.4) is simple and a measurement of the intensity of light scattered by the sample can be used for calculating the concentration c of the substance to be determined, provided that the scattering index (i.e. the constant a) is known. The value of a can be determined by calibration with a standard sample. Tabulating the a values, analogously to molar absorptivities in absorption spectrophotometry, for solutions of pure substances in a given solvent, is difficult because the measured values of the intensity of the scattered radiation are dependent on the instrument used for registering the spectrum, but for compilation purposes should be independent of it. This can be achieved either by measuring the absolute intensity of the scattered radiation, or by measuring the intensity relative to a generally accepted standard. Measurement of the absolute intensity is too difficult to achieve in normal laboratory practice as it involves a number of corrections which are subject to large errors. The use of an intensity standard should allow tabulation of reliable values of the scattering indices for a substance in a given medium. Cyclohexane and carbon tetrachloride have been suggested as solvents which may also be used as standards for measuring the relative scattering index of a dissolved substance.

6.2 SOME COMMENTS ON THE ABSORPTION METHODS

Transmittance, T, is the quantity most commonly measured in absorption spectrophotometry, and is expressed as a ratio of intensities:

$$T = \frac{I}{I_0} \tag{6.5}$$

where I is the intensity of the light transmitted by the sample and I_0 the intensity of the light transmitted by a reference sample (usually containing as undetectable amount of the absorbing species). It is assumed that the cells used are matched with regard to their reflection and absorption characteristics, or that corrections can be applied. Because of the exponential nature of the absorption, the relation used for quantitative work is

$$A = \log \frac{1}{T} = \log \frac{I_0}{I} = kcl \tag{6.6}$$

the familiar Lambert–Beer law. The absorbance A is a linear function of the concentration c of the substance to be determined, which gives the number of absorbing centres in unit volume and layer thickness l. The coefficient k is characteristic of the substance to be determined and is known as the *absorptivity* or *absorption coefficient.* In principle the absorbance for a given species and concentration should be independent of the spectrophotometer used, since it is based on the ratio of the measured intensities. In practice, variation in spectral band-width and stray-light characteristics from one instrument to another can result in differences in the apparent absorption coefficients. Fortunately, the intensities I and I_0 used in absorption methods usually differ by less than two orders of magnitude, so the differences in standard transmittance measurements made with different instruments are comparatively small. In Raman spectrometry, however, the intensity I_0 of the exciting beam is several orders of magnitude greater that that of the Raman scattering intensity I, and measurement of I/I_0 is subject to comparatively large error, even when the difficulties involved in building the spectrometer are overcome.

6.3 DIFFICULTIES AND PROCEDURE IN QUANTITATIVE ANALYSIS

Quantitative analysis based on Raman spectra involves the use of *standards* (see Section 6.4). These may be a pure sample of the substance to be determined, or a solution (mixture) of one or more substances to be determined and a suitable background substance (matrix). The spectra are recorded under identical experimental conditions for the standards and the samples to be analysed. It is here that the major difficulty arises, viz. the maintenance of identical illumination conditions for standards and samples. For liquids, rectangular glass cells can be used (see Section 3.3.2), which often ensures sufficient reproducibility of illumination. For powders and samples where scattering takes place on the surface, it is essential to use an *internal standard,* i.e. a suitable substance added in constant concentration to standards and samples to be analysed. The intensity of the Raman band of the substance to be determined is measured relative to the intensity of the band of the internal standard.

The analytical procedure includes the following steps.

1. Identification of the sample components from the Raman spectrum.
2. Selection of suitably strong analytical bands which do not overlap other bands of the sample.
3. Choice, if necessary, of an internal standard with a strong band lying close to the analytical band of the sample but not overlapping it.
4. Preparation of standards of composition approximating that of the sample.
5. Measurement of the integrated or peak intensity of the analytical bands (cf. Section 8.6) for the standards and sample under identical experimental conditions.
6. Plotting the analytical curves (dependence of analytical band intensity on concentration of the component to be determined) for the particular components.
7. Determining the concentration of the particular components with the aid of the analytical curves.

6.4 EXAMPLES OF DETERMINATIONS

The examples described below have been selected so as to include materials in different states of aggregation, i.e. gas, liquid and solid samples, and compounds from a variety of chemical classes.* The examples presented also illustrate typical advantages and disadvantages of methods of quantitative analysis based on Raman spectra.

A common advantage of analytical methods based on infrared and Raman spectrometry is the possibility of identifying structural elements within a molecule and determining their concentration. Laser Raman spectrometry has the additional advantages that it requires only very small samples and that it is non-destructive.

For gases and for liquid mixtures of similar compounds, e.g. petroleum fractions containing liquid hydrocarbons, quantitative analysis can be done by using standards which are pure samples of the hydrocarbons to be determined. In this case we can assume that the scattering index a [Eq. (6.4)] of the pure substance is identical, within experimental error, to that of the same substance in the environment of the mixture being analysed.

---

* The examples have been chosen from the work done at the Institute of Industrial Chemistry in Warsaw by H. Barańska, A. Łabudzińska and W. Andrzejewska.

Reproducible results of intensity measurements can only be obtained when suitable measuring cells are used.

In single-beam spectrometers, where the intensity of the scattered band is not measured relative to the intensity of the laser reference band, which varies with time because of variations in intensity of the exciting laser beam, it is advisable to check the stability of the excitation conditions. This can be achieved by using a so-called *external standard*, e.g. by measuring at suitable intervals the intensity of the 802 cm^{-1} band of pure cyclohexane (always contained in the same cell, placed in the 'sample' position in the spectrometer).

6.4.1 Determination of butene isomers in a gas sample

The spectra of the four butene isomers are presented in Fig. 6.1, and it can be seen that they differ considerably. For 2-methylpropene and *trans*-but-2-ene there are distinct, well-separated bands at 805 cm^{-1} and 503 cm^{-1} and these are useful for determining the two components in mixtures by the pure standard method. The spectra can be recorded by placing the sample in an ordinary cylindrical cell at atmospheric pressure. The cell

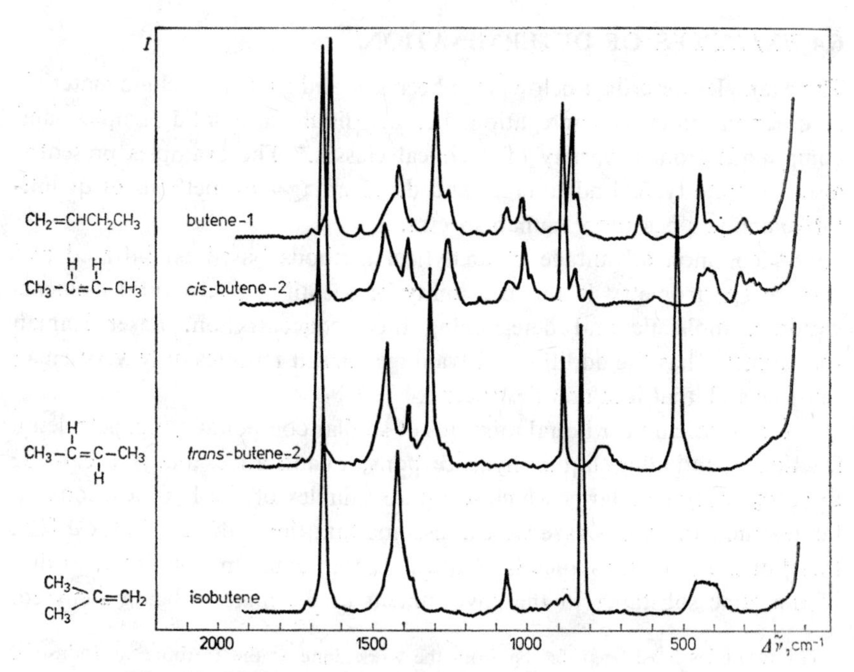

Fig. 6.1—Raman spectra of butene isomers in the gas phase.

is provided with a valve and after the air has been removed with a rotary pump, the cell is filled with the sample gas from a suitable plastic balloon at atmospheric pressure. The spectra of the pure isomers and the mixture to be analysed are recorded under identical measuring conditions (slit-width, photomultiplier voltage, amplification, recorder sensitivity) within the analytical band range. The background is eliminated by drawing a straight line parallel to the base line between the chosen wave numbers (always the same for the given analytical band) (between $\tilde{\nu}_1$ and $\tilde{\nu}_2$ in Fig. 8.5). The peak intensity is in this case (in view of the constancy of the band-width at half height, see Section 8.6) a sufficiently accurate measure of the band intensity. The concentration x of the component to be determined in the mixture is calculated (in % v/v) by using Eq. (6.4):

$$x = \frac{100 I_x}{I_{100}} \%$$

where I_x and I_{100} are the peak intensities of the same analytical band for the analysed mixture and the pure standard, respectively.

6.4.2 Determination of sulphate in aqueous solutions of ammonium sulphate

The samples of ammonium sulphate solution sometimes contained nitric acid or ammonia. Rectangular glass cells were used and the standard solutions had a composition similar to that of the samples to be determined. For aqueous ammonium sulphate solutions with no additives, or containing an excess of ammonia, the sensitivity was high and superior to that normally found for the Raman scattering method, viz. a limit of detection of 0.05% SO_4^{2-}. This is because of the very high scattering coefficient of the totally symmetric vibration of the tetrahedral SO_4^{2-} anion, with a band at 982 cm^{-1}. The presence of a sufficiently high concentration of nitric acid produced a very strong lowering of the intensity of that band (see Fig. 6.2), owing to the shift in the equilibrium $H^+ + SO_4^{2-} \rightleftharpoons HSO_4^-$ ($pK_2 = 2.0$ for sulphuric acid). The HSO_4^- band lies at 1048 cm^{-1} and overlaps with the strong NO_3^- band at 1050 cm^{-1}. This example provides a good illustration of the need to match the composition of the standards to that of the sample when the medium interacts strongly with the species determined. It is possible to determine SO_4^{2-} even in solutions containing high concentrations of nitric acid by using the appropriate analytical curve (Fig. 6.2, curve 2), but the sensitivity of determination is greatly decreased. To avoid the inconvenience of making matrix-matched standards, the standard-additions method could be used.

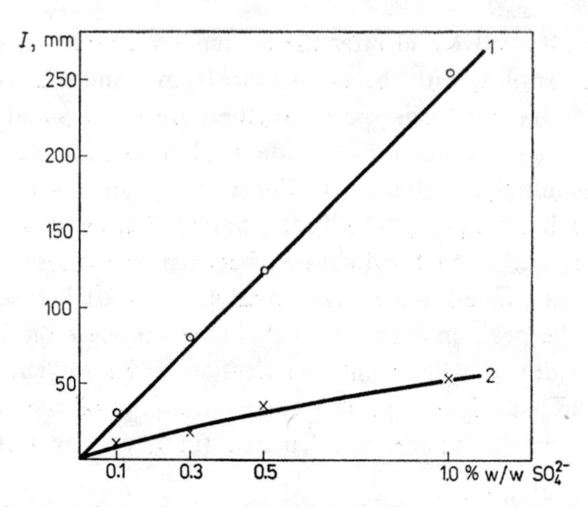

Fig. 6.2—The analytical curves used in the determination of the SO_4^{2-} ion (band at 982 cm^{-1}). I is the peak intensity of the analytical band (mm). (1) Aqueous solutions of $(NH_4)_2SO_4$. (2) Aqueous solutions of $(NH_4)_2SO_4$ containing a constant amount of HNO_3.

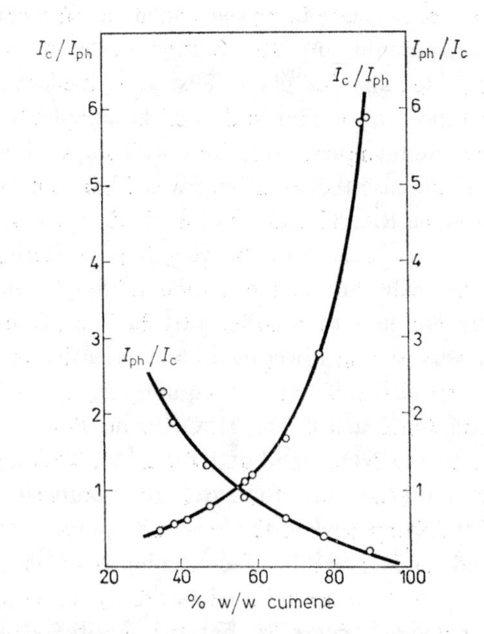

Fig. 6.3—Analytical curves for cumene–phenol mixtures: I_{ph} and I_c are the peak intensities of the analytical bands for phenol at 818 cm^{-1} and cumene at 745 cm^{-1}, respectively.

6.4.3 Analysis of a two-component mixture

If a mixture consists of only two components, these can act as internal standards for each other and glass capillaries can be used as the measuring cells. Measurements of the intensity of the analytical band of the given components in the same sample, in different capillary cells, will not give reproducible results, since the capillaries will differ in wall thickness and cross-section. It is impossible to illuminate them with the laser beam and collect the scattered radiation in an identical way. The effect of using different cells and hence different illuminations can be eliminated by measuring the analytical band intensity ratios of the components, since the individual intensities are influenced to the same extent by these two factors. The analytical curves plotted on the basis of measurements made for standard solutions are presented in Fig. 6.3 for a cumene–phenol mixture [14]. The percentage of cumene in the sample can be found from the ratio I_{ph}/I_c or I_c/I_{ph} for the sample.

From the slope of the curves, it can be seen that the error in the determination of cumene content is smaller when the steeper parts of the curves are used. Therefore the choice of which ratio to use depends on the sample composition. The analytical curves are not symmetrical because of the different scattering indices of the selected analytical bands for phenol and cumene ($a_{ph} \neq a_c$). The results obtained by this method were in agreement with those from infrared spectrophotometry.

Capillaries are convenient measuring cells, since they require only very small amounts of sample, which can be sealed in the capillary and hence kept unchanged for long periods of time. This procedure has also proved very convenient for analysis of coloured or opalescent samples.

6.4.4 Analysis of five-component aromatic hydrocarbon fractions

The mixtures under examination contained *o*-, *m*- and *p*-xylene, ethylbenzene and toluene. The determinations were performed with mixed-standard solutions of compositions close to those of the samples. A rectangular glass cell was used. Analytical curves were plotted from the averages of five successive recordings of the spectrum in the analytical-band range, and the peak heights of the analytical bands were used as the measure of the intensities.

Some samples had a yellowish shade and absorbed part of the incident radiation (488.0 nm line of the argon-ion laser) and also some of the scattered radiation. Determination of the components of such samples on the basis of analytical curves derived from measurements on completely colourless mixed solutions, would lead to significant errors. The problem was solved by adding to the samples and standards a constant amount of cyclohexane as an internal standard. On this basis the correction for the absorption of radiation in the sample could be calculated. The correction factor used was the ratio of the intensities at the 802 cm^{-1} cyclohexane band maximum for the colourless mixed standard and for the coloured sample. In Fig. 6.4 sections of the spectrum, including the cyclohexane

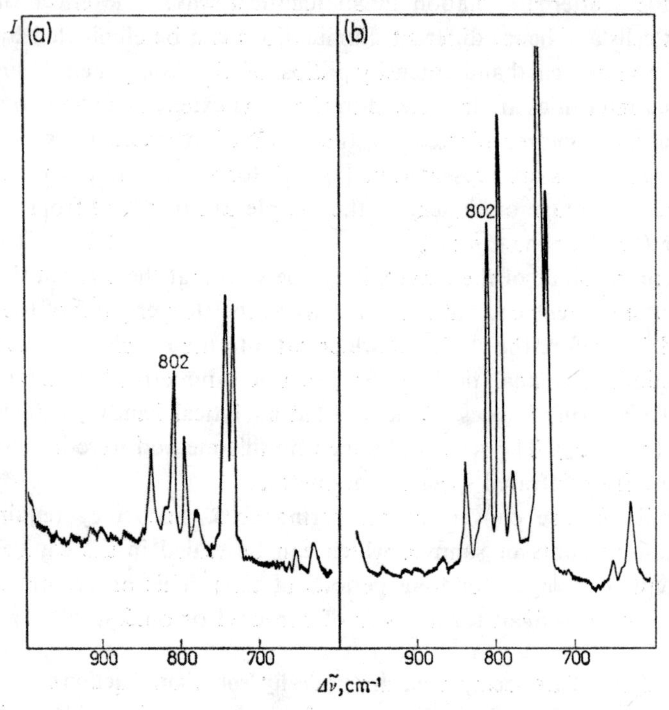

Fig. 6.4—Raman spectra of a yellowish sample (a) and of the colourless mixed standard (b) (mixed aromatic hydrocarbons). The sample and standard contained 12% v/v of cyclohexane as internal standard.

band, are presented for the colourless mixed standard and the yellowish sample, the amount of cyclohexane added being 12% v/v. The height of the 802 cm^{-1} band is distinctly smaller for the coloured sample, though the cyclohexane contents and measuring conditions are identical in both cases. When the amount of cyclohexane added to the coloured sample and colourless standard was varied in the range 4–12% v/v, the value of the correction factor remained constant within experimental error.

6.4.5 Determination of poly(vinyl chloride) in powder samples [14]

In this study an inorganic compound was used as the internal standard. Despite differences in the texture of the poly(vinyl chloride) powder (PVC) and that of the inorganic salt, this standard was shown to be satisfactory.

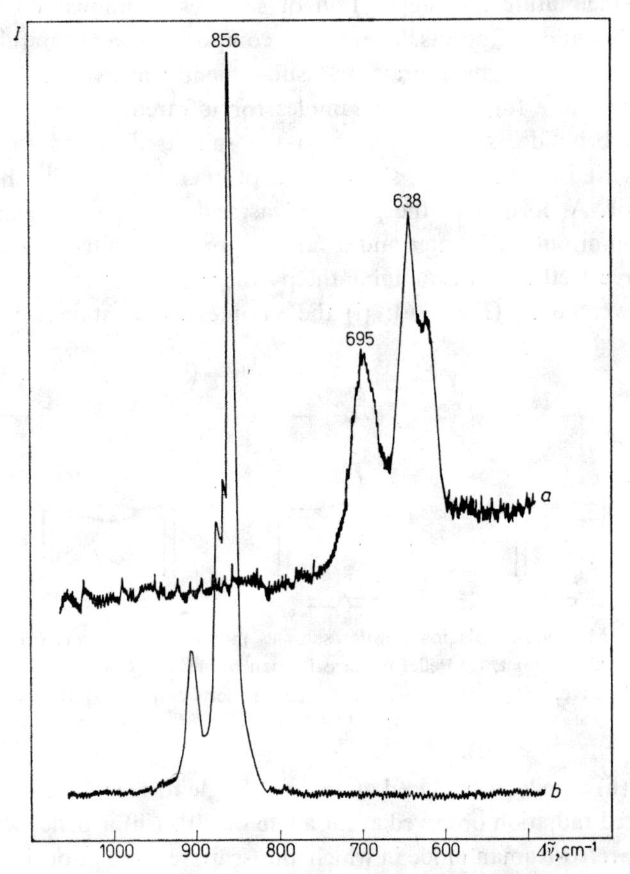

Fig. 6.5—Sections of the Raman spectra of poly(vinyl chloride) (*a*) and potassium barium chromate (*b*) containing the chosen analytical bands.

The content of poly(vinyl chloride) which could be determined in the mixture ranged from 10 to 90%v/v. Potassium barium chromate, which exhibits a band at 856 cm^{-1}, lying close to the poly(vinyl chloride) analytical bands at 638 and 695 cm^{-1} (C—Cl stretching frequency) was chosen as the internal standard. Sections of the spectra are presented in Fig. 6.5. The contents of the internal standard in both samples and standards must be the same;

in our case it was 5%v/v. The advantage of using inorganic salts containing complex anions showing high symmetry is the occurrence of strong bands corresponding to totally symmetric vibrations of the anion. The few remaining bands are usually much weaker. Potassium bromide was used as a carrier, facilitating the preparation of samples of known composition and the fabrication of pellets. The sample components were carefully mixed by grinding in an agate mortar and subsequently in a mill. This is the routine procedure for preparing samples for infrared spectrophotometry. Potassium bromide is hygroscopic, so the salt itself (dried to constant weight) as well as the pellets, should be kept in closed weighing bottles in a desiccator. A portion of the powder was sealed in a glass capillary of about 1.3 mm outer diameter and a further portion was used to make the pellet. Three methods of examining the powder sample in the Raman spectrometer were used (Fig. 6.6): (i) the sample was contained in a glass

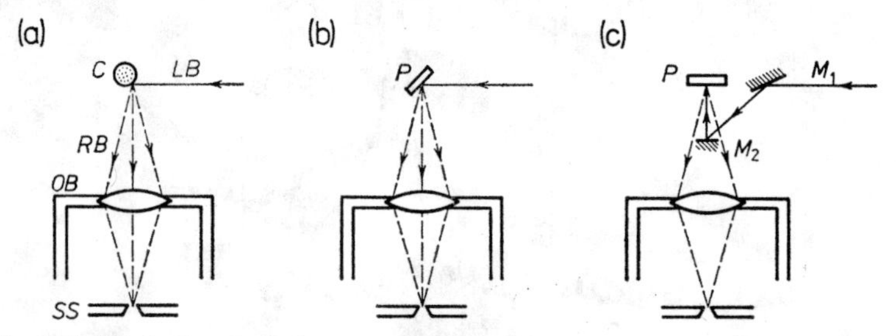

Fig. 6.6—Methods of placing powder samples in the spectrometer. (a) Powder n capillary. (b) Pellet. (c) Pellet in retro-Raman mode. (*C*) Capillary. (*P*) Pellet. (*OB*) Objective. (*SS*) Spectrometer slit. (*LB*) Laser beam. (*RB*) Raman beam. (M_1, M_2) Mirrors.

capillary, (ii) a pellet was placed at an acute angle to the incident beam and the scattered radiation observed at an angle of 90°, (iii) a pellet was examined in the retro-Raman mode in which the scattered radiation is observed at an angle of 180° to the incident beam.

It was found that there was a linear relationship between the intensity of the PVC analytical band relative to that of the chromate reference band and the PVC content in the sample (Fig. 6.7). The peak heights of the analytical and reference bands were used as a measure of intensity. The analytical bands were recorded five times in succession and the averages of the five peak intensities were used in the calculations.

Analysis of the standard deviation of the method has shown that sample heterogeneity is the greatest source of error; the second most important factor is the error in the measuring technique, but instability of the instrumentation can be neglected. The best results were obtained with the pellets when the scattered radiation was observed at right angles. Use of the capillary significantly facilitated the determination. The relative error did not exceed 5% in any of the procedures used. To reduce the effect of heterogeneity of the sample it is recommended to use a rotating cell, or at least to change the point of illumination of the sample before each measurement in the series used for averaging.

I638/I856
a
b
c
0.8
0.6
0.4
0.2
0
20 40 60 80
% w/w PVC

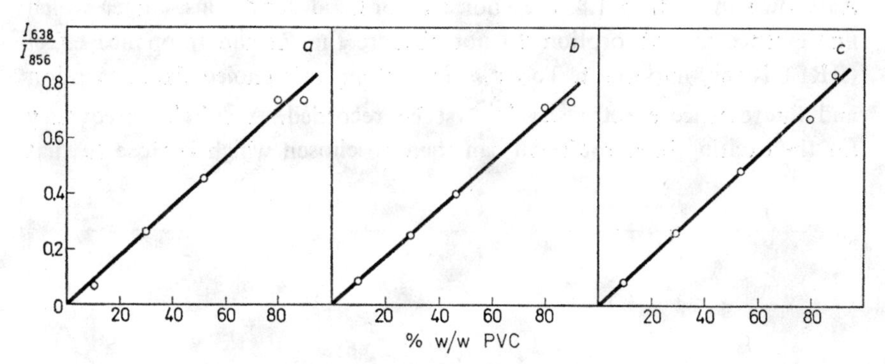

Fig. 6.7—Analytical curves for poly(vinyl chloride) (PVC) obtained by the three measuring techniques presented in Fig. 6.6

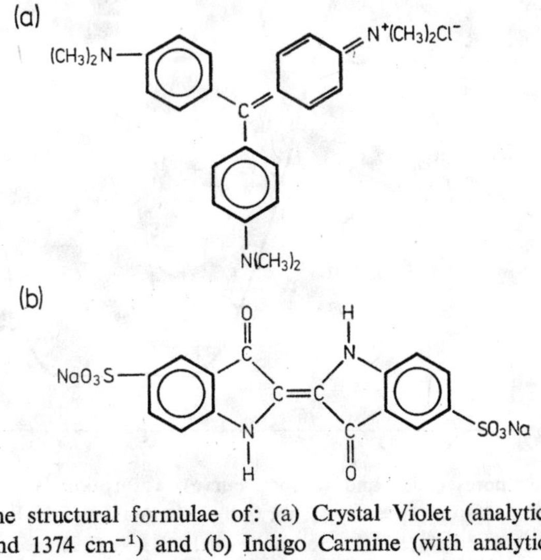

(a)
(CH3)2N
N+(CH3)2Cl−
C
N(CH3)2
(b)
NaO3S
SO3Na
O
H
N
C
C=C

Fig. 6.8—The structural formulae of: (a) Crystal Violet (analytical bands at 917, 1182 and 1374 cm^{-1}) and (b) Indigo Carmine (with analytical bands at 1246, 1298 and 1358 cm^{-1}).

6.4.6 Determination of small amounts of organic dyes in aqueous solution by the Raman resonance effect

The Raman resonance effect makes it possible to obtain a sensitivity some 3 or 4 orders of magnitude greater than that of conventional Raman spectrometry. Determination of organic substances (which absorb in the visible region) in small concentrations is thus possible.

This effect has been used to determine Crystal Violet and Indigo Carmine (Fig. 6.8) in aqueous solution at concentrations of 10^{-5}–10^{-4} *M*. As shown in Section 1.8, the choice of optimal conditions, under which fluorescence and absorption do not obscure the Raman resonance effect (RRE), is very important. To assist in making this choice the absorption and fluorescence spectra should first be recorded. A suitable frequency for the exciting laser radiation can then be chosen which is close to that

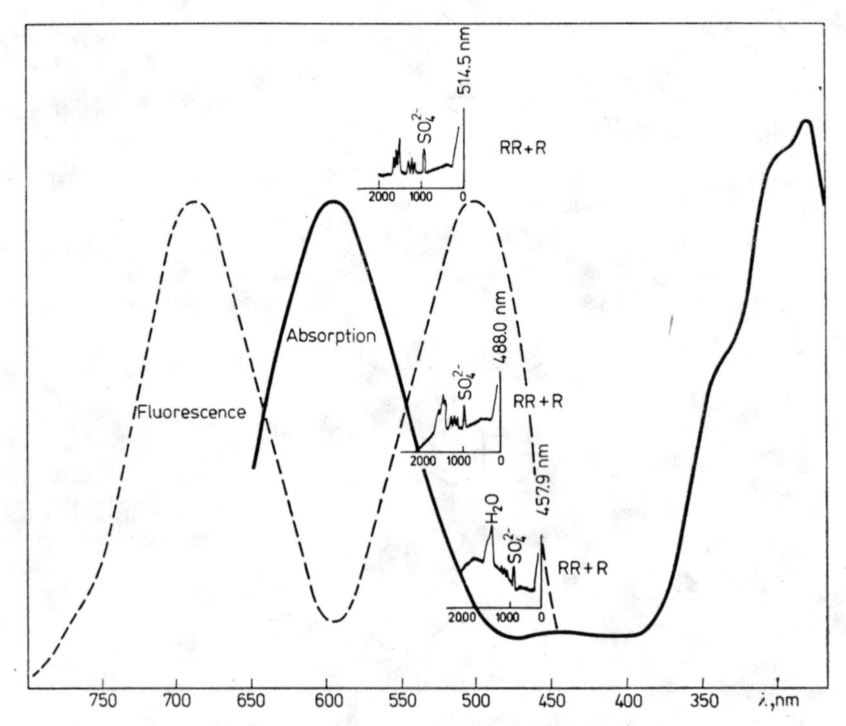

Fig. 6.9—The fluorescence band (dashed curve), absorption band (solid line) and the resonance Raman spectra (RR) of Indigo Carmine, and the Raman spectra (R) of SO_4^{2-} ions and water. The abscissae of the RR and R spectra are scaled in cm^{-1}.

of the absorption maximum, but distant from the nearest fluorescence band maximum. The next step is the choice of an internal standard. In the determinations considered here we chose sodium sulphate because of the strong Raman band due to the totally symmetric SO_4^{2-} vibration at 982

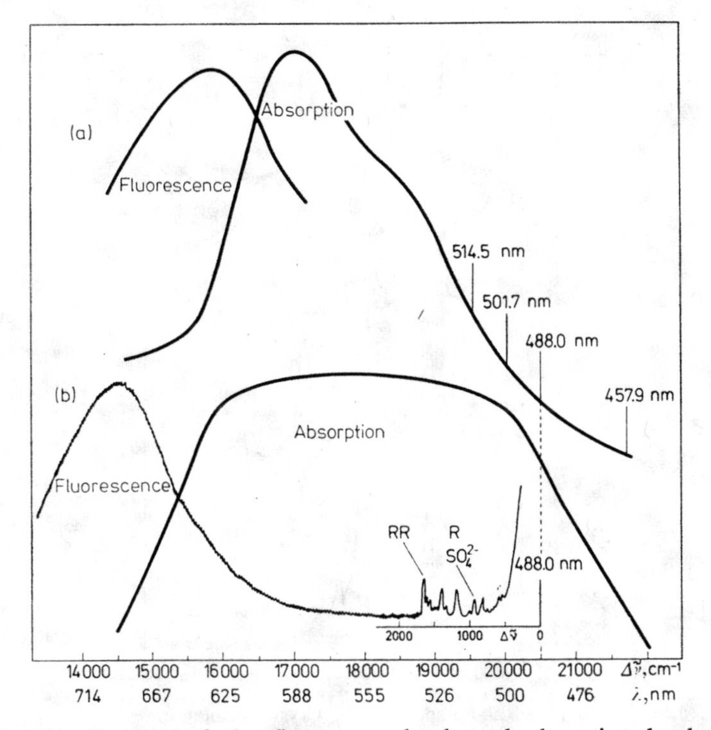

Fig. 6.10—Contours of the fluorescence bands and absorption bands of Crystal Violet. (a) In aqueous solution. (b) In the form of a pellet with KBr. (RR). The resonance Raman spectrum of Crystal Violet in pellet form excited by the 488.0 nm line of the Ar^+ laser (R). Raman band of SO_4^{2-} ions (Na_2SO_4 was added as internal standard).

cm^{-1}. Figure 6.9 illustrates the spectra for an aqueous solution of $10^{-4}M$ Indigo Carmine containing sodium sulphate as the internal standard. The Raman spectra presented were obtained by using three different wavelengths: 457.9, 488.0 and 514.5 nm of the argon-ion laser. They therefore occupy different positions with respect to the absorption and fluorescence band maxima. The most satisfactory well-resolved spectrum was that obtained with an excitation wavelength of 514.5 nm. This spectrum has

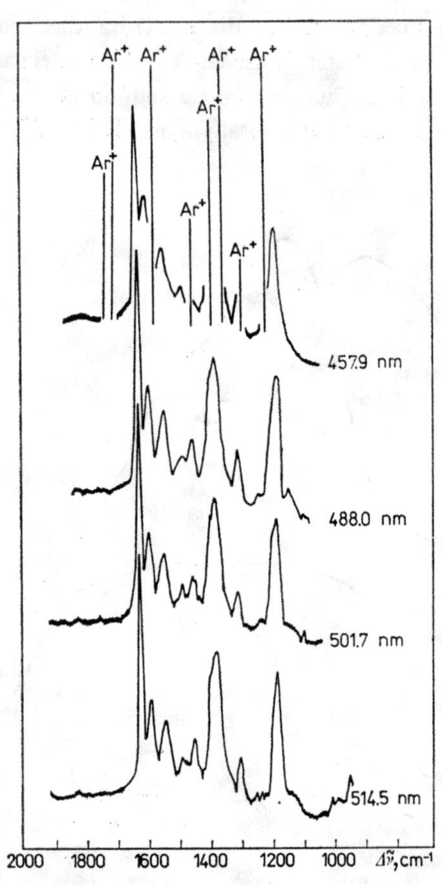

Fig. 6.11—Resonance Raman spectrum of Crystal Violet in pellet form with KBr, excited by an Ar^+ laser beam at various wavelengths (in the case of excitation by the 457.9 nm line, the non-laser Ar^+ lines are marked by vertical lines). The intensity of the various laser lines and the conditions of recording were the same. The same pellet was analysed without changing its position in the spectrometer.

the lowest fluorescence background and greatest intensity of the RRE bands. Owing to the re-absorption of the scattered Raman radiation by the sample there is an advantageous reduction in the intensity of the band at 1640 cm^{-1} due to water. In the other two spectra this band overlaps the Indigo Carmine band. The difference in the contours of the absorption and fluorescence bands for Crystal Violet in aqueous solution and in a potassium bromide pellet is illustrated in Fig. 6.10. No definite changes of band intensity in the RRE spectrum with variation of the excitation line wavelength were observed for the KBr pellet (Fig. 6.11) but distinct changes were observed for the aqueous solution (Fig. 6.12).

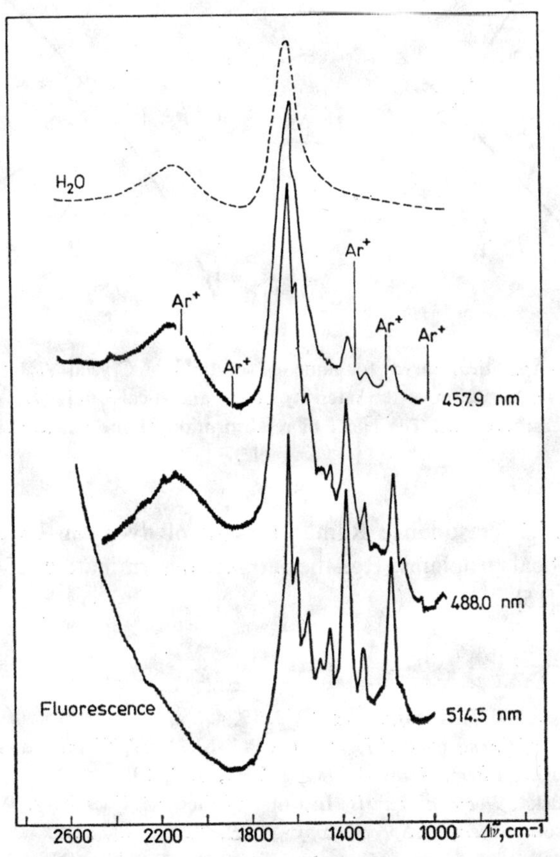

Fig. 6.12—Resonance Raman spectrum of Crystal Violet in aqueous solution, excited by an Ar^+ laser beam at various wavelengths. The intensity of the various laser lines and the conditions of recording were the same. The liquid sample was placed in a rotating cylindrical glass cell.

Owing to the use of an internal standard (added in constant amount to each solution), the analytical curves for relative analytical band intensity vs. concentration of substance being determined were monotonic (Fig. 6.13). This allows analysis with a 3–5% relative error in the average results. The measurements were performed by using the rotary cell described in

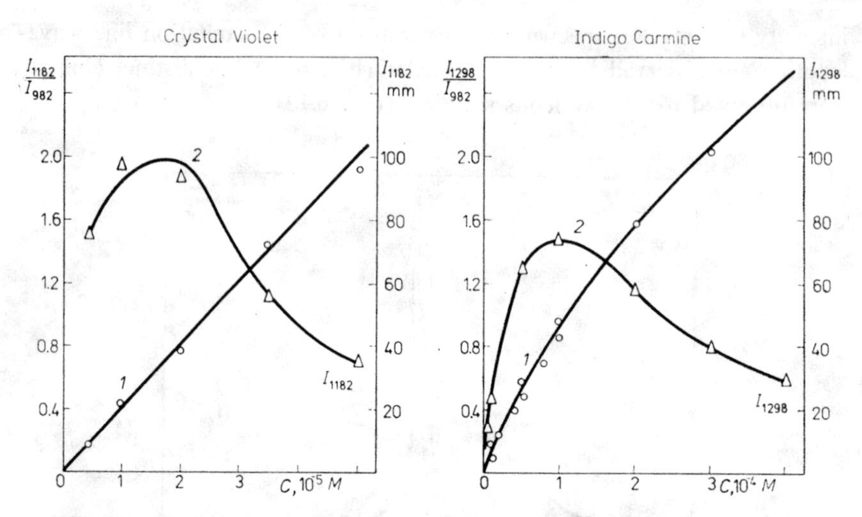

Fig. 6.13—Analytical curves for aqueous solutions of Crystal Violet and Indigo Carmine (*1*). Variation of the intensity of the analytical bands (I_{1182} and I_{1298}) with concentration (*2*). The effect of re-absorption of the scattered radiation is visible.

Section 1.8. The resonance Raman spectra of dyes can be used to solve other analytical problems, e.g. the indirect determination of oxidant concentrations [15].

REFERENCES

[1] P. Szynagel and Z. Kęcki, *Nafta*, 1963, **19**, 45.
[2] *Handbook of Chemistry and Physics* (in Polish), WNT, Warszawa, 1961, p. 1611.
[3] H. Barańska, *Chem. Anal.* (*Warsaw*), 1963, **8**, 3, 13.
[4] H. Barańska, *Doctoral Thesis*, Institute of General Chemistry, Warsaw, 1968.
[5] W. R. Brode, *Chemical Spectroscopy*, Wiley, New York, 1945.
[6] M. G. Mellon, *Analytical Absorption Spectroscopy*, Wiley, New York, 1950.
[7] *Absorption Spectrometry* (in Polish), J. Świętosławska (ed.), PWN, Warszawa, 1962.
[8] G. W. Ewing, *Instrumental Methods of Chemical Analysis*, McGraw-Hill, New York, 1960.
[9] H. J. Bernstein, in *Proc. 5th European Congress on Molecular Spectroscopy*, Amsterdam, 1960.
[10] J. A. Koningstein, *Doctoral Thesis*, University of Amsterdam, 1962.
[11] Z. Kęcki, *D. Sc. Thesis*, Warsaw University, Warsaw, 1961.
[12] L. A. Woodward and J. H. B. George, *Proc. Phys. Soc.*, 1951, **64**, 780.
[13] L. Onsager, *J. Am. Chem. Soc.*, 1936, **58**, 1486.
[14] H. Barańska and A. Łabudzińska, *Chem. Anal.* (*Warsaw*), 1966, **21**, 93.
[15] H. Barańska, A. Łabudzińska and J. Młodecka, *Applied Spectrosc.*, 1983, **37**, 476.

7

Selected applications of laser Raman spectroscopy

H. Barańska and A. Łabudzińska

This chapter describes briefly some applications of laser Raman spectroscopy. This technique has in recent times been used widely in analytical practice and, in the authors' opinion, may develop further in this field.

7.1 POLYMERS AND BIOPOLYMERS

Since lasers began to be used as excitation sources in Raman spectroscopy, this method has been developing rapidly as a tool for examining polymers and biological substances. The low power of the mercury-arc exciting beam and the fluorescence due to the mercury-arc radiation, which occurred frequently in polymer spectra, resulted in serious difficulties in the recording of the spectra. Nowadays, the Raman scattering and infrared absorption spectra of polymer samples are used as important sources of information about the vibrations of macromolecules and are thus useful for identifying the polymer structure [1–6].

7.1.1 Experimental techniques

Raman scattering spectra can be produced from samples in different states of matter. Solid samples are examined as powders, fibres or pieces having various shapes, e.g. cylinders, cuboids or irregular lumps. The sample may vary from totally amorphous material, through mixtures of different phases to single crystals. Fibres and foils, which are stretched during the process of manufacture, contain polymer chains which are orientated to different degrees.

Polymers can be spectroscopically examined in the liquid state in the form of molten samples or as solutions.

The investigation of polymers by Raman spectroscopy involves recording of the spectra, measurements of band polarization and a study of the variation in the spectra with the angle between the incident laser beam and the orientation of the polymer chains. Coloured biological substances or polymers can be investigated by resonance Raman spectroscopy (cf. Section 1.8).

The principal difficulty that attends the recording of Raman spectra is the fluorescence, which frequently occurs with polymers and biopolymers. This often arises from small amounts of fluorescent impurities, which are very difficult to remove. Sometimes the polymer itself shows fluorescence; in those cases the procedure suggested in Section 3.3.1 may be used.

Depolarization ratios can be measured only for completely transparent samples, and these may be difficult to obtain for some polymers.

7.1.2 Information on the structure of polymers from Raman spectra

A Raman spectrum, which is a vibrational spectrum, mirrors all the molecular features responsible for the actual vibrations which the molecule undergoes.

Raman scattering depends on the polymer chain structure, so the bands appearing in the spectrum are associated with vibrations of the functional groups of the macromolecule. The more chemically complex the molecule (e.g. a copolymer), i.e. the greater the number of functional groups of different types contained within a segment of the polymer chain, the more complex the spectrum and the more difficult the identification of the bands.

Another property reflected in the spectrum of a macromolecule is its stereoregularity. Thus Raman spectra are useful for defining the polymer tacticity.

In a macromolecule, the intramolecular vibrations are affected by the polymer chain conformation. They depend on whether or not the chain is in the form of a zig-zag, a regular spiral or a random coil. The study of band-drifts in Raman spectra may be useful for defining the chain conformation as well as for observing chain transformations, for example, from a spiral to a coil form.

In the spectra of crystalline polymers there are bands associated with crystal lattice vibrations, which are useful in structural analysis.

Resonance Raman spectra of coloured polymers may provide information on polymerization processes and the structure of polymerization products.

This brief summary shows that Raman spectroscopy may be used for investigating the structure of a macromolecule segment, identifying the chain conformation and orientation, as well as defining its position in the unit cell. It has proved particularly useful in some cases to combine both methods of vibrational spectroscopy, viz. Raman scattering and infrared absorption, since the spectra are complementary. The strongest bands observed in Raman spectra are associated with skeletal vibrations of the polymer chain, whereas in infrared absorption spectra, bands assigned to the vibrations of side groups are dominant.

Raman spectroscopy, up to the present day, has not been used in the routine analysis of polymers, but rapid progress in this field indicates a future increase in its popularity for this purpose.

7.1.3 Information on the structure of biopolymers from Raman spectra

For natural polymers (biopolymers), the Raman bands may be attributed to the same macromolecular properties as those described in the preceding section. Biopolymers, however, are chemically more complex than most other polymers. A molecule of a biopolymer can be compared with that of a multicomponent copolymer, since the chain carries many different functional groups. These groups give rise to numerous bands in vibrational spectra. The chains of biopolymer molecules are commonly in the form of helices and are cross-linked, especially in the case of proteins. These factors give rise to complex Raman spectra, which are difficult to interpret. Therefore only seldom can all the bands in a biopolymer spectrum be assigned to particular molecular vibrations. The assignment of bands is easier when both the Raman and infrared spectra are available.

In spite of these difficulties, Raman spectroscopy is now more commonly employed for examining biopolymers, since it supplies information which is difficult to obtain by other methods.

Raman spectroscopy has made possible the study of phenomena of great importance for biological processes, for instance, intra- and intermolecular interactions, bipolymer chain conformation, and structural changes due to variation of temperature and pH. From Raman spectra the type of bonds formed between a biopolymer molecule (e.g. a protein) and a small coloured species (ketone, vitamin, metal-ion etc.) can be deduced. The possibility of using water as a solvent in Raman spectroscopy greatly facilitates the investigation.

The normal Raman effect is used to study proteins, nucleic acids, lipids and fatty acids, examined in the form of solutions, suspensions and liquid crystals. A study of the structural arrangement of the phospholipid alkyl chain [7] is given here as an example. From the position and intensity of the bands associated with the chain C—C stretching mode, deductions were made concerning synclinal and antiperiplanar conformations. The bands attributed to the CH_2 deformation mode supplied information on changes of the chain conformation and on interchain interactions. The bands attributed to the C—H stretching mode provided information on intrachain and interchain interactions. The observed changes in band intensity with increase of temperature were assigned to changes of chain conformation, which took place in the course of melting of multi-lamellar aggregates.

From Raman spectra of nucleic acids, deductions were made concerning the helical configuration of the chain.

The position and intensity of the C—C and C—N bands were used as a basis for investigation of the chain conformation in a protein molecule. The C—S and S—S groups give rise to very strong Raman bands, whereas the corresponding infrared absorption bands are usually very weak. Thus Raman spectra are more useful for investigating the disulphide bridges in proteins.

Raman spectra are employed in the examination of intermolecular interactions and the formation of complexes of proteins with nucleic acids (viruses, chromatin, nucleosomes).

The study of changes in conformation and of interactions during the transition from a solid polymer to its aqueous solution, provides essential information about biological processes. As previously stated, vibrational spectra of aqueous solutions can be examined only by Raman scattering, because of the strong absorption of infrared radiation by water.

Raman spectra of crystalline proteins, and specially prepared powders or films, have been used in the investigation of vibrations of the entire molecule.

The resonance Raman effect is often used in the study of biopolymers, since it is possible to observe interactions between some small coloured molecules (chromophores) and a biopolymer molecule (e.g. a protein). The method is applicable to natural dyes and coloured substances, such as carotenes, chlorophyll, vitamin B_{12}, porphyrins containing Fe(II) and protein molecules bonded to metals. In addition, compounds which have been specially 'labelled' with coloured substances to take advantage of the resonance Raman effect can be investigated. Proteins and nucleic acids which absorb radiation in the ultraviolet region have also been studied by using the resonance Raman effect.

From resonance Raman spectra deductions can be made concerning the nature of the bond formed between a small coloured molecule and a protein. It is also possible to observe the changes in structure which take place when the bond is being formed. The method has been employed to study the displacement of the iron atom from haemoglobin and the formation of a bond between a dye and the albumin in ox-blood serum. It has also been used to investigate the bonds being formed in immune response [5, 6]. The main advantage of resonance Raman spectroscopy is that the scattering is many orders of magnitude more intense than normal Raman scattering. Thus the method requires only a low dye concentration, which is of fundamental importance to researchers into biological problems.

7.2 COMPLEX COMPOUNDS

Ionic [8] and molecular complexes [9] have always been subjects of great interest to the spectroscopist. Knowledge of the structure of complexes and the strength of chemical bonds linking the two components of the complex, contributes to an understanding of the chemical changes which occur in the course of some reactions (including catalytic reactions important in chemical synthesis). The production of complexes by the formation of hydrogen bonds is considered to be an important process in living organisms. Interactions in solution, particularly aqueous solutions of electrolytes [10, 11] have already been investigated spectroscopically by using mercury-arc Raman spectrometers but laser excitation sources have greatly increased the possibilities in this field.

Many scientists working on molecular complexes find it difficult to decide on a precise interpretation of this term. According to Mulliken and Person [9], if a strictly defined number of components are associated and the forces bonding the components together are slightly stronger than the van der Waals forces, the association can be spoken of as a *molecular complex*. Although much attention has been devoted to this problem, we do not yet know precisely the type of forces which act between the components of a molecular complex. Theoretical studies, spectroscopic investigations of the component structures, mechanism of formation and calculations made of equilibrium and thermodynamic constants have all contributed to a widening of our understanding of the problem.

The number of Raman bands and their depolarization ratios have often been used as basic experimental data for identifying a complex and defining its structure. The relative intensities of particular bands, which are proportional to the concentrations of the corresponding components forming the complex, can be used in the determination of equilibrium constants. The thermodynamic constants of the reaction forming the complex can be determined from the change in relative band intensity with temperature.

Complexes are commonly examined as solutions in an appropriate solvent. The solution is placed in a liquid-sample cell (cf. Section 3.3.2), and laser Raman spectrometry used to determine (cf. Chapter 6) the structure and to measure the band depolarization ratio. If a molecular complex is insoluble or unstable in solution, powder samples can be used (cf. Section 3.3.2). In this case, however, difficulties arise in measuring the band intensity and measurement of the band depolarization ratio is impossible.

7.3 MATRIX ISOLATION SPECTROSCOPY

Infrared spectroscopic studies involving the matrix isolation technique have been known since 1950 and have proved very useful for the examination of reactive molecules and free radicals. Later, the technique was used for the investigation of intermolecular interactions, such as the formation of dimers, trimers and oligomers. The possibility of defining precise conditions for the formation of different oligomers that is offered by the use of this modern technique, allowed scientists to obtain fundamental information on intermolecular forces. At present, the matrix isolation technique is also being used in the structural analysis of condensed gas mixtures.

The *matrix isolation technique* can also be employed in laser Raman spectroscopy, since laser excitation sources provide ideal conditions for the recording of vibrational–rotational spectra (cf. Sections 3.1.3 and 3.3). The technique [12–15] consists in rapidly cooling a gas mixture so as to form the matrix material in which the investigated species are preserved. This results in the trapping, in a frozen matrix, of the molecules to be examined. Owing to their chemical inertness, the noble gases and nitrogen are generally used as matrix materials. Other gases, e.g. O_2, CO, CO_2, SF_6 and $Si(CH_3)_4$ have been employed in specific studies. The rapid cooling, which produces the rigidity of the matrix cage and prevents diffusion and reactions between molecules of the trapped species, requires the use of very low temperatures, for example, that of liquid nitrogen or, better, liquid helium. The required temperature is usually defined as $\frac{1}{2}T_m$, where T_m is the absolute temperature of the melting point of the gas used. Advances in the construction of cryostats, which nowadays are provided with closed helium systems and high-vacuum facilities, have led to a wider use of the matrix isolation technique in current research.

A *cryostat* suitable for Raman spectroscopy in which the species under examination are trapped in a frozen matrix is shown schematically in Fig. 7.1. The cryostat is connected to a high-vacuum line. The end of

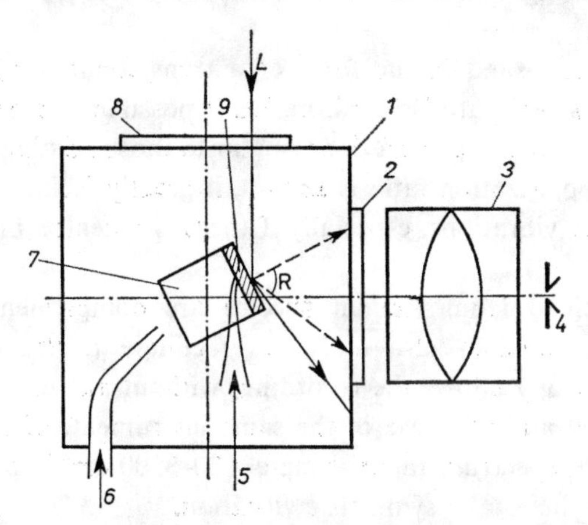

Fig. 7.1—Schematic diagram of an arrangement for Raman spectroscopic studies using the matrix isolation technique. (*1*) Cryostat. (*2* and *8*) Flat, parallel glass windows. (*3*) Converging lens. (*4*) Raman spectrometer entrance slit. (*5*) Thermocouple. (*6*) Inlet for mixture of matrix gas and sample vapour. (*7*) Metal block cooled to the temperature of liquid nitrogen or helium. (*9*) Deposition of the examined species in the solidified matrix gas. (*L*) Laser beam. (R) Scattered radiation.

a metal block (*7* in the diagram, made of aluminium, copper, or platinized copper), is cooled by liquid nitrogen or liquid helium in the cryostat. The block can be rotated around an axis perpendicular to the plane of the diagram. A mixture of the matrix gas and the vapour of the species under examination issues from a pipe, *6*, and is deposited as a thin film on the cold polished metal surface of the block. When the thickness of this film is judged sufficient, the gas supply is cut off and the block is rotated through an angle of 20°, so that the laser beam, *L*, illuminates the matrix surface. The beam reflected from the surface does not reach the window, *2*, through which only the scattered radiation R is allowed to pass. This radiation is focused on the entrance slit of the Raman spectrometer, *4*, by the lens system, *3*.

Molecules of the species under investigation are isolated from each other by the matrix material. The extent of isolation depends on the molar ratio of the species being examined and the matrix gas and so the desired experimental conditions can be established by adjusting this ratio.

The matrix method can be used in the production of free radicals, or for initiating chemical reactions with a single intense flash of ultraviolet light. The resulting chemical changes can be observed in the recorded Raman spectra.

A matrix deposited in the form of a transparent film, in which the molecules of a sample are isolated, makes it possible to measure the band depolarization ratio under conditions similar to those existing in the gaseous state. If the depolarization ratio is known, it greatly facilitates the analysis of the molecular vibrations, especially if there is a centre of symmetry in the molecule.

Data obtained from Raman spectra are complementary to those provided by infrared spectra and thus facilitate a structural analysis. Raman spectroscopy allows the recording, without any change in cryostat window material and with use of the same instrument, of the total vibrational–rotational spectrum (approximately 30–3500 cm^{-1}) and it is always easy to identify the totally symmetric vibration.

Raman spectroscopy often permits the differentiation of *cis* and *trans* isomers and the identification of a macromolecular conformation or a symmetry group, when other methods provide contradictory evidence. Nibler [15] has provided examples of the different experimental results obtained for LaF_3 and PrF_3 by infrared spectroscopic methods and the electron diffraction technique. The problem was only solved by using the Raman matrix isolation technique. From the position, intensity and depolarization ratio of Raman bands, the symmetry group of LaF_3 was identified as C_{3v} (pyramidal structure) and that of PrF_3 as D_{3h} (planar structure). These results were confirmed by electron-density mapping.

REFERENCES

[1] T. R. Gilson and P. J. Hendra, *Laser Raman Spectroscopy*, Wiley-Interscience, New York, 1970.

[2] M. C. Tobin, *Laser Raman Spectroscopy*, Wiley-Interscience, New York, 1971.

[3] *Infrared and Raman Spectroscopy of Biological Molecules*, T. M. Theophanides (ed.), Reidel, Dordrecht, 1979.

[4] A. T. Tu, *Raman Spectroscopy in Biology*, Wiley-Interscience, New York, 1982.

[5] P. R. Carey, *Resonance Raman Labels—Probes for Drug, Immunochemical and Enzyme Reactions*, in *Proceedings Fifth International Conference on Raman Spectroscopy*, Schulz Verlag, Freiburg, 1976.

[6] P. R. Carey, *Resonance Raman Spectra of Biological Materials: Approaches to Interpretations*, in *Proceedings Sixth International Conference on Raman Spectroscopy*, Heyden, London, 1978.

[7] G. J. Thomas Ir., *Recent Applications of the Normal Raman Effect for the Study of Biologically Important Molecules*, in *Proceedings Sixth International Conference on Raman Spectroscopy*, Heyden, London, 1978.

[8] D. E. Irish, *Raman Spectroscopy of Complex Ions in Solution*, in *Raman Spectroscopy*, H. A. Szymański (ed.), Plenum Press, New York, 1967.

[9] J. Yarwood (ed.), *Spectroscopy and Structure of Molecular Complexes*, Plenum Press, New York, 1973, p. 2.

[10] Z. Kęcki, *D. Sc. Thesis*, Warsaw University, Warsaw, 1961.

[11] Z. Kęcki, W. Kołos, W. Libuś, S. Minc and L. Stolarczyk, *Spectroscopic Investigations of the Structure of Electrolyte Solutions* (in Polish), PWN, Warszawa, 1969.

[12] K. Szczepaniak, *Matrix Isolation Techniques in Infrared Spectroscopy*, in *Vibrational Spectroscopy Methods* (in Polish), L. Sobczyk (ed.), PWN, Warszawa, 1979.

[13] H. E. Hallam, *Matrix Isolation Techniques*, in *Laboratory Methods in Infrared Spectroscopy*, R. G. J. Miller and B. C. Stace (eds.), Heyden, London, 1972.

[14] H. E. Hallam, *Vibrational Spectroscopy of Trapped Species*, Wiley, London, 1973.

[15] J. W. Nibler, *Raman Matrix Isolation Spectroscopy*, in *Advances in Raman Spectroscopy*, J. P. Mathieu (ed.), Vol. 1, Heyden, London, 1973.

8

Fundamental concepts in spectroscopy

H. Barańska

In recent years a number of books have been published [1–6] which provide a good introduction to molecular spectroscopy, so this chapter will deal only with the fundamental concepts of molecular spectroscopy which have been used in the body of the book.

8.1 QUANTITIES CHARACTERISTIC OF AN ELECTROMAGNETIC WAVE

In the methods of molecular spectroscopy the electromagnetic wave plays the part of a transmitter of information about the structure of molecules and their number in a given sample.

The *electromagnetic wave* can be described as a periodical propagating perturbation of the electric and magnetic fields where the electric vector is perpendicular to the magnetic vector and both are in the plane perpendicular to the direction of wave propagation.

The electromagnetic wave is characterized by *wavelength* or the *frequency of vibration*, i.e. the changes in the electric or magnetic vector taking place in the corresponding fields.

These quantities are connected by the relation:

$$\lambda = \frac{c}{\nu} \tag{8.1}$$

where c is a constant, the velocity of the electromagnetic wave in a vacuum.

The length of the electromagnetic wave, λ, is usually expressed in nm. The frequency, ν, is expressed as the number of vibrations per second (Hz). It can also be characterized by the wavenumber $\tilde{\nu}$, (cm^{-1}). The following relation exists between wavenumber and frequency:

$$\tilde{\nu} = \frac{\nu}{c} \tag{8.2}$$

Equations (8.1) and (8.2) imply the relationship between the wavelength expressed in cm and the wavenumber:

$$\tilde{\nu} = \frac{1}{\lambda}\ [\mathrm{cm}^{-1}] \tag{8.3}$$

Different wavelength ranges used in studying the properties of molecules involve the application of different spectroscopic methods. Table 8.1 gives the numerical data characterizing some of these.

8.2 ENERGY LEVELS, POPULATION RATIO, DEGENERACY

The electromagnetic wave is also characterized by the photons which it carries (cf. Section 1.1). The amount of quantized energy supplied by the photons determines the type of excitation of the molecule interacting with the electromagnetic wave. The molecule can undergo energy transitions involving rotations, vibrations or changes in electronic energy levels (Table 8.1). The quantized energy levels of a molecule are presented diagrammatically in Fig. 8.1. Transitions between the rotational levels result in rotational spectra (observed for gases by Raman spectroscopy and as absorption in the far-infrared or microwave region). Transitions between the vibrational levels are accompanied by rotational transitions. Transitions between the electronic energy levels are accompanied by vibrational and rotational transitions.

Any assembly of molecules contains molecules with different degrees of excitation, i.e. in different energy levels. The number of molecules in a given energy state depends on the temperature, and is expressed by the *Boltzmann distribution function*. The population ratio of two different energy states is given by the equation:

$$\frac{N_i}{N_j} = \exp[-(E_i - E_j)/kT] \tag{8.4}$$

Table 8.1—Comparison of spectrometric methods and corresponding radiation ranges*

Method	Energy transition type	Photon energy, eV	λ	ν, Hz	$\tilde{\nu}$, cm^{-1}
Absorption spectrophotometry in the ultraviolet region (UV spectrometry)	electronic	12–3	100–400 nm	3×10^{15}–7.5×10^{14}	10^5–2.5×10^4
Absorption spectrophotometry in the visible region (Vis spectrometry); Fluorimetry	electronic	3–1.8	400–700 nm	7.5×10^{14}–4×10^{14}	2.5×10^4–1.4×10^4
Absorption spectrophotometry in the infrared region (IR spectrometry); Raman spectrometry†	vibration -rotation	5×10^{-1}–10^{-2}	2.5–10^2 μm	1.2×10^{14}–3×10^{12}	4×10^3–10^2
	pure rotational	10^{-2}–10^{-3}	10^2–10^3 μm	3×10^{12}–3×10^{11}	10^2–10

* Ranges assigned are not precisely determined and numerical data are only approximate.

† In the case of Raman spectrometry the numerical values given correspond to Raman shifts ($\Delta\tilde{\nu}$ cm^{-1}).

where N_i and N_j are the numbers of molecules in the higher energy state E_i and lower energy state E_j, respectively, k is the Boltzmann constant and T the absolute temperature.

In the case of degeneracy of an energy level it is necessary to take into account the degree of degeneracy of that level in calculating the population ratio. *Degeneracy of a level* occurs when several wave functions Ψ correspond to its energy.

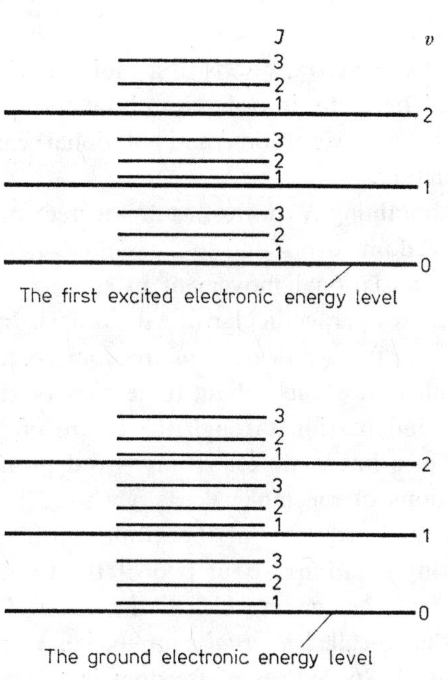

Fig. 8.1—Schematic diagram of the electronic energy, vibrational energy (v) and rotational energy (J) levels of a diatomic molecule.

The time-independent Schrödinger equation permits us to calculate the energy and position of a given particle. The solution of the equation uniquely determines the possible energy levels of the particle. The position of the particle, on the other hand, is given as the probability of its occurrence at a given point in space. This probability is determined by the square of the absolute value of the wave function, Ψ^2, (or by the product $\Psi\Psi^*$ in the case of conjugate functions), the wave function being the solution of the Schrödinger equation for the given case. It happens for some energy levels that a particle can have several different wave functions Ψ, i.e. several kinds of motion of the particle are connected with the level. This is precisely the case for degeneracy.

Degeneracy of normal modes (cf. Section 8.3) occurs when different vibrational modes correspond to the same wave frequency, resulting from the same energy of transition between their vibrational levels.

8.3 DEGREES OF FREEDOM AND NORMAL MODES OF VIBRATION

The possibilities of energy transitions in a molecule are closely connected with its structure. Therefore, information about those transitions encoded in the spectra (electronic, vibrational and rotational) can be used in studying molecular structure.

A molecule containing N atoms has $3N$ degrees of freedom. Three of these degrees of freedom correspond to *translations* (shifts) of the molecule as a whole (any translational movement in space can be represented as a shift along three axes perpendicular to each other). In the case of a non-linear molecule three further degrees of freedom correspond to the *rotations* of the molecule as a whole—about three axes of rotation perpendicular to each other and passing through the centre of gravity of the molecule. In the case of a linear molecule only two degrees of freedom correspond to the rotations of the molecule as a whole. This results from the fact that the spatial coordinates of the atoms forming the molecule are changed only during rotations about two of the axes. The third axis of rotation coincides with the line on which all the atoms of the linear molecule are located. For the so-called *internal motions*, i.e. vibrations of the molecule, there remain $3N-6$ degrees of freedom in the case of a non-linear molecule and $3N-5$ in the case of a linear molecule. In this way the number of characteristic vibrations for a given molecule can be calculated. These vibrations are called *normal modes* and are described by normal coordinates. Each mode is described by one normal coordinate Q. A normal mode is defined as a vibration in which all the atoms of the molecule vibrate with the same frequency and pass simultaneously (in the same phase) through the position of equilibrium.

During normal vibrations of a molecule its atoms undergo displacements along the bonds or displacements changing the bond angles. According to the kind of change the modes are called:

(a) *stretching modes* or *valence modes* when there is a change in the bond lengths,

(b) *bending* or *deformational modes* when there is a change in the bond angles.

With regard to the symmetry of the displacements, *modes* are classified as *symmetric* or *antisymmetric.*

In the case of planar molecules or molecular fragments, the modes in which the displacements occur in the plane are called *in-plane modes* and those in which the displacements bring the atoms above and below the plane are called *out-of-plane modes.*

In more complex molecules the vibrations are complicated and consequently have various names. According to the various kinds of motion of the atoms, *scissoring, rocking, wagging* and *twisting modes* are distinguished. They are all types of deformational vibration.

Figure 8.2 shows, as an example, the vibrations of the acetylene molecule and some vibrations of the dichloromethane molecule, together with the commonly used mode type symbols. Totally symmetric skeletal (*breathing*) modes of cyclic compounds are of interest in Raman spectroscopy because they are recorded as very intense totally polarized bands.

In every normal mode all the atoms of the molecule are involved. However, to certain groupings of atoms in the molecule, e.g. to functional groups or to fragments of the carbon skeleton, we can assign certain vibration frequencies (called *group frequencies*), which change relatively little for various molecules containing the same groups. This follows from the fact that the vibration amplitudes of different atoms in a molecule vary considerably. The value of the vibration amplitude has a decisive effect on the intensity (or the absorbance) of the recorded band. Thus, some vibrations can be characteristic for a given group of atoms in the molecule because the remaining atoms vibrate with very low relative amplitudes during that vibration.

Frequencies connected with a given kind of vibration are determined by general rules:

(a) the frequencies of stretching vibrations are higher than those of deformational vibrations of the same molecule or group of atoms;
(b) the vibration frequency assigned to the X—H group (where H is the hydrogen atom and X is another atom) is usually higher than the vibration of the corresponding X—Y group;
(c) the frequency of a stretching vibration depends on the strength of the chemical bond (on its order). The stretching vibration frequency of the carbon–carbon bond increases from the single bond C—C to the double bond C=C and the triple bond C≡C.

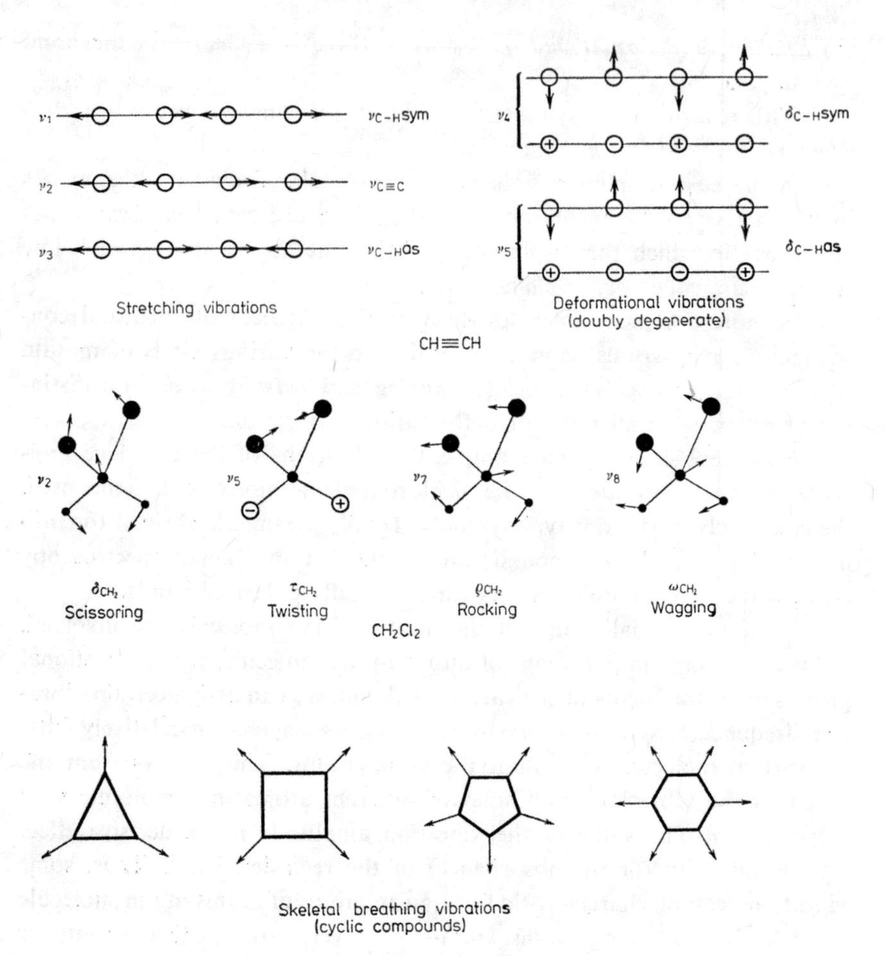

Fig. 8.2—Different types of normal vibration modes.

8.4 ROTATIONAL LEVELS

The energies of quantized rotational levels of a molecule are calculated by modelling its movement as a rigid rotor and by the quantization of the angular momentum.

The equation representing the rotational energy expressed in cm^{-1} has the form:

$$E_j = \frac{h}{8\pi^2 Ic} J(J+1) \tag{8.5}$$

In this equation h is Planck's constant, c is the velocity of light, J is the quantum number of rotation ($J = 0, 1, 2 \ldots$), and I is the moment of inertia given by $I = \sum_i m_i r_i^2$ (where m_i is the mass of a given atom of the molecule and r_i its distance from the axis of rotation passing through the centre of gravity of the molecule).

In molecular spectroscopy the term $h/8\pi^2 Ic$ appearing in Eq. (8.5) is denoted by the symbol B:

$$B = \frac{h}{8\pi^2 Ic} \text{ cm}^{-1}$$

The interval between the lines in the rotational spectrum of a linear molecule is constant and equal to $2B$ cm^{-1}.

From the experimentally determined values of B for the molecule in which the atoms have been substituted by different isotopes, we can obtain the necessary number of equations to calculate the moments of inertia I and hence the interatomic distances in the molecule. Thus, the rotational spectra permit us to calculate important structural parameters of the molecule, such as bond lengths and bond angles.

Line intensity in a rotational spectrum depends on the population of the rotational energy states of the molecule. This population is expressed in terms of the Boltzmann distribution:

$$N_J = (2J+1)N_0 \exp[-J(J+1)h^2/8\pi^2 IkT] \tag{8.6}$$

where N_J is the number of molecules in the level with rotational quantum number J at absolute temperature T; N_0 is the number of molecules in the ground level (unperturbed, $J = 0$); h is Planck's constant and I is the moment of inertia of the molecule.

In fact the molecule does not behave as a rigid rotor and that is why in exact calculations we introduce corrections for the rotation–vibration coupling and for the centrifugal force.

8.5 VIBRATIONAL LEVELS

The energies of quantized vibrational levels of a molecule are calculated by assuming the harmonic oscillator as a model of its vibrations.

From the quantum theory of the harmonic oscillator we obtain the equation of its eigenvalues, i.e. the energy values corresponding to the consecutive energy levels determined by the vibrational quantum numbers ($v = 0, 1, 2 \ldots$). These values are the so-called *stationary levels* which have definite life times, i.e. definite periods during which a molecule remains in a given energy level. The vibrational energy of a given level is expressed by the equation:

$$E = h\nu_0(v+\tfrac{1}{2}) \tag{8.7}$$

where ν_0 is the *eigenfrequency of the harmonic oscillator*. This frequency can be found from an equation which is a classical expression for the vibration frequency of the harmonic oscillator:

$$\nu_0 = \frac{1}{2\pi}\sqrt{\frac{f}{m_r}} \tag{8.8}$$

where f is the force constant and m_r is the reduced mass. In the case of a diatomic molecule $m_r = m_1 m_2/(m_1+m_2)$, where m_1 and m_2 are the masses of atoms 1 and 2, respectively.

It follows from Eq. (8.7) that the vibrational energy can never be zero and that the transition between two neighbouring energy levels of consecutive quantum numbers v and $v+1$ corresponds to an energy difference $\Delta E = E_{v+1} - E_v = h\nu_0$.

In fact a molecular vibrator does not behave as a harmonic oscillator. The force acting on the atoms of a vibrating molecule is not proportional

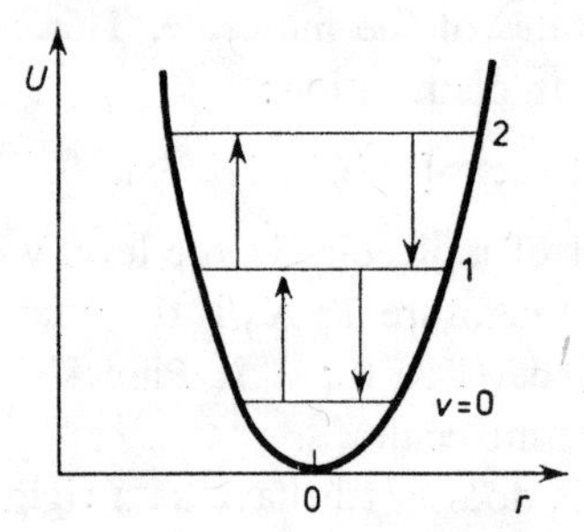

Fig. 8.3—Potential energy (U) curve of a harmonic oscillator; r is the displacement from the state of equilibrium. Possible vibrational transitions are marked with arrows.

to their displacements from the position of equilibrium as it is in the case of a harmonic oscillator. The molecular vibrator behaves as an anharmonic oscillator. The potential energy curve of a diatomic harmonic oscillator is a parabola (Fig. 8.3) and that of an anharmonic oscillator is the curve of the Morse function (Fig. 8.4).

The selection rule defining the energy transition allowed in the case of a harmonic oscillator does not provide an accurate description of the

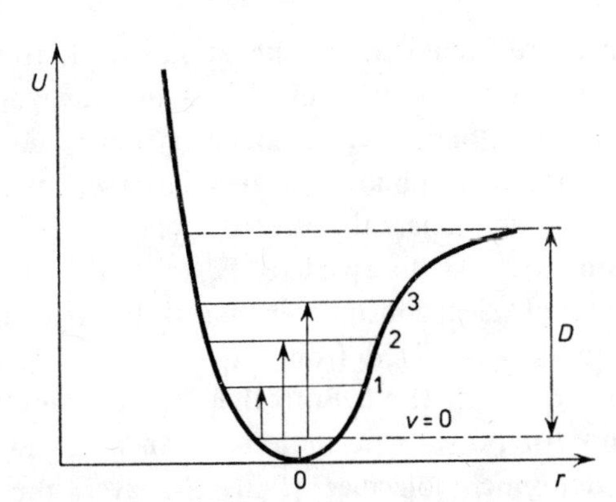

Fig. 8.4—Potential energy (U) curve of an anharmonic oscillator; r is the displacement from the state of equilibrium and D is the dissociation energy of the molecule. The transitions corresponding to $\Delta v = +1; +2; +3; \ldots$ are marked with arrows.

transitions actually observed for a molecule. They can be described more closely in terms of the *selection rule* for the *anharmonic oscillator*.

The selection rule for the harmonic oscillator is

$$\Delta v = \pm 1 \tag{8.9}$$

so that the only possible transitions are those between neighbouring energy levels. They are the so-called *fundamentals*.

The analogous selection rule for the anharmonic oscillator has the form:

$$\Delta v = \pm 1, \pm 2, \pm 3, \ldots \tag{8.10}$$

so that in addition to the fundamental $\Delta v = \pm 1$ there are transitions $\Delta v = \pm 2$ or $\pm 3, \ldots, \pm n$. They are the so-called *overtones*.

Figures 8.3 and 8.4 show the potential energy curves and the energy transitions possible in the case of harmonic and anharmonic vibrations respectively. To each normal vibration mode of the molecule we can assign a single vibrator, characteristic for that vibration mode. A molecule of $3N-6$ degrees of freedom is represented by $3N-6$ vibrators. In the vibrational spectra of actual polyatomic molecules, in addition to the fundamentals and the overtones, we also observe *combination tones*; their frequencies are the sums or the differences of the frequencies of the fundamentals or the overtones, or of the overtones and the fundamentals of different vibrators.

It is also possible to observe the phenomenon of *Fermi resonance* between coupled molecular vibrators. The conditions required for this type of resonance are identical, or almost identical, frequencies of the fundamental tone of one vibrator and of an overtone or a combination tone of the other vibrator and an identical symmetry for the vibrations to which these tones correspond. We then observe, instead of the band corresponding to the fundamental tone, two bands of different frequencies. This phenomenon was first observed by Fermi for the CO_2 molecule.

A fundamental physical constant which is connected with the molecular structure and can be calculated from experimental spectra is the force constant f appearing in Eq. (8.8). Both the formulae and the calculations of force constants for polyatomic molecules are much more complicated than those for diatomic molecules [1–3]. However, the data obtained from infrared and Raman spectra of molecules which have been isotopically enriched and the use of modern computing techniques have made it possible to determine the force constants of chemical bonds in polyatomic molecules.

The intensity of individual bands in the spectrum of a compound is related to the probability of transition between the corresponding vibrational levels and to the populations of those levels, which in turn depends on the absolute temperature. Dependence on absolute temperature is given by the Boltzmann function of energy distribution. For vibrational transitions it has the form:

$$N_v = N_0 \exp[-(E_v - E_0)/kT] \tag{8.11}$$

where N_v is the number of molecules in the vibrational level v; N_0 is the number of molecules in the lowest vibrational level (the fundamental $v = 0$), k is the Boltzmann constant, T is the absolute temperature and E_v and E_0 are the energies of the vibrational levels characterized by the vibrational quantum numbers v and 0, respectively.

It follows from Eq. (8.11) that in any assembly of molecules the most heavily populated is the fundamental level [1] and hence the most conspicuous transition in the vibrational spectrum is the transition $0 \rightarrow 1$. The transitions between higher levels, e.g. $1 \rightarrow 2$, $2 \rightarrow 3$ (the so-called *hot bands*), only appear at sufficiently high temperatures, at which the populations of the higher levels are suitably increased.

8.6 INFRARED AND RAMAN SPECTRA

The measurement of absorption in the infrared region (infrared spectrometry) and of Raman scattering (Raman spectrometry) makes possible the recording of the vibration–rotation spectrum of a molecule.

The *selection rules*, i.e. the conditions necessary for the rotation or vibration of the molecule to appear in the form of a rotational line or a vibration–rotation band, differ for the two kinds of spectrometry. The selection rules are listed in Table 8.2. The differences in the relative intensities of the same bands in the infrared and Raman spectra have been collected in Table 8.3.

One condition which determines whether or not a molecular vibration will appear as a band in an infrared or Raman spectrum is the efficiency of the relaxation process, i.e. the return of the molecule to its original energy state (the state before the absorption of a quantum of energy or interaction with that quantum). Having absorbed the quantum of energy the molecule passes to a higher energy state. If the molecules were unable to lose the absorbed energy, saturation would ensue. The recording of the radiation absorbed or scattered by the sample, which requires an appreciable time, could not then be performed by any instrument. The return to the original state occurs either by the emission of a quantum of absorbed energy or by its conversion into thermal (kinetic) energy during intermolecular collisions.

Infrared and *Raman spectra* can be considered as graphs of the absorbance (or transmittance) or of the intensity of scattered radiation, respectively, vs. the frequency of the recorded radiation, usually expressed as the wavenumber $\tilde{\nu}$ (cm^{-1}). The spectra consist of a number of bands corresponding to the normal modes of the molecule (their fundamentals and overtones, and also combinations and Fermi resonances in the case of polyatomic molecules).

A *band* is characterized by the following parameters (Fig. 8.5):

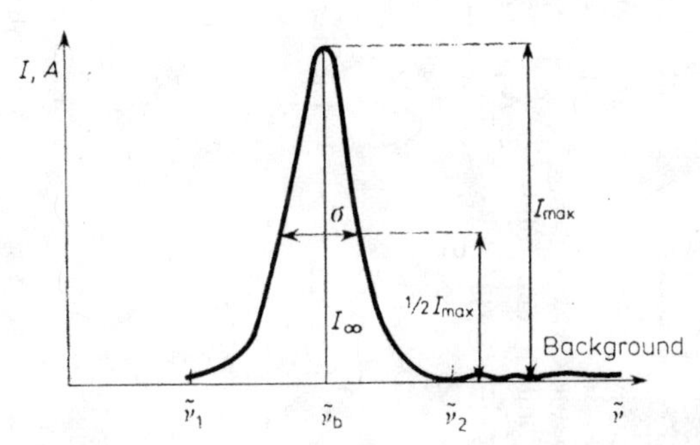

Fig. 8.5—Shape and parameters of a band. (I) Intensity. (A) Absorbance). ($\tilde{\nu}$) Wavenumber (cm^{-1}). ($\tilde{\nu}_b$) Wavenumber of the band. (σ) Half-width. (I_{max}) Maximum intensity. (I_∞) Integrated intensity.

Table 8.2—Selection rules of rotational and vibrational transitions in infrared and Raman spectrometry

Type of transition	Infrared	Raman
Rotations (observed in gases only; in liquids and solids they are almost completely damped)	1. The energy of the photon matches the energy difference of rotational levels: $h\nu = \Delta E_{rot}$ 2. A permanent dipole moment is necessary 3. Transitions occur between neighbouring rotational levels: $\Delta J = 0, \pm 1$ (ΔJ—change in the rotational quantum number)	1. The energy difference of the incident and scattered photon matches the energy difference of rotational levels: $h\nu_0 - h\nu_s = \Delta E_{rot}$ 2. A permanent dipole moment is unnecessary—can be recorded rotational spectra of homonuclear gases, e.g. N_2, O_2 3. $\Delta J = 0, \pm 1, \pm 2$
Vibrations	1. The energy of the photon matches the energy difference of vibrational levels: $h\nu = \Delta E_{vib}$ 2. The dipole moment changes during vibration: $\left(\frac{\partial \mu}{\partial Q}\right)_0 \neq 0$ 3. $\Delta v = +1, +2, +3, \ldots$ (Δv is the change in the vibrational quantum number)	1. The energy difference of the incident and scattered photon matches the energy difference of vibrational levels: $h\nu_0 - h\nu_s = \Delta E_{vib}$ 2. There is a change in polarizability during vibration: $\left(\frac{\partial \alpha}{\partial Q}\right)_0 \neq 0$ 3. $\Delta v = \pm 1, \pm 2, \pm 3, \ldots$ (transitions for $\Delta v = \pm 2, \pm 3, \ldots$, i.e. overtones are considerably less conspicuous than in IR)

Table 8.3—Band intensities in the vibration–rotation infrared and Raman spectra

Group	IR (relative absorbance)	R (relative intensity)
Polar groups with high permanent dipole moment, e.g. OH, NH, CO	very strong	weak
C—S S—S Si—O—Si	medium	strong or very strong
C=C C≡C	strong	very strong
Skeletal vibrations	medium	strong
Totally symmetric vibrations, particularly of aromatic and alicyclic rings	very low or zero (forbidden in IR)	strong or very strong

(a) band maximum position—the wavenumber of the band ($\tilde{\nu}_b$);
(b) band intensity (absorbance)—either the maximum (I_{max} or A_{max}) or the integrated value (I_∞ or A_∞);
(c) band half-width (σ).

The *position of the band* (the corresponding frequency or wavenumber) is a fundamental parameter of qualitative structural analysis. *Band intensity* is a fundamental parameter of quantitative analysis. The *integrated intensity of a band* in a Raman spectrum is given by the equation:

$$I_\infty = \int_{-\infty}^{+\infty} I(\tilde{\nu})\mathrm{d}\tilde{\nu} \tag{8.12}$$

where $I(\tilde{\nu})$ is the function expressing the dependence of band intensity I on wavenumber $\tilde{\nu}$. *Integrated absorbance* is defined analogously, the intensity I in Eq. (8.12) being replaced by the absorbance A. In practice both integrated intensity and absorbance are determined by measuring the area between the band profile and the background by use of a planimeter. The *background* is usually defined by choosing two points on the band profile at fixed wavenumber values for the given band, according to its shape ($\tilde{\nu}_1$ and $\tilde{\nu}_2$ in Fig. 8.5) and joining them with a straight line. In the case of

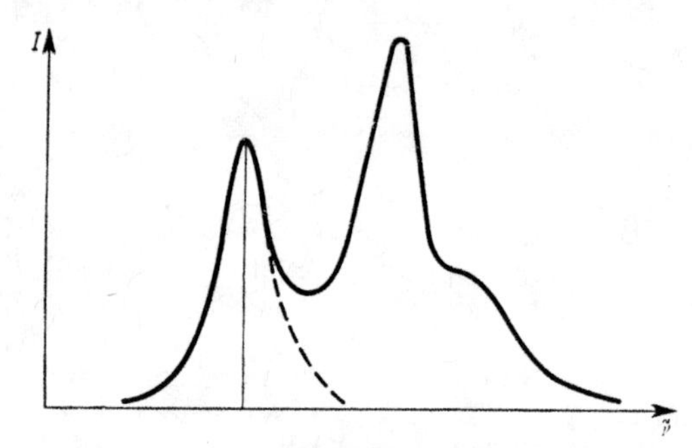

Fig. 8.6—Overlapping bands.

overlapping bands each individual band can be separated by assuming that its profile is symmetrical (Fig. 8.6).

In quantitative analysis, we compare the intensities of the band for the sample and that for the standard, recorded under the same experimental conditions. A sufficiently reliable measure of integrated intensity is the intensity at the band maximum (I_{max}) provided that the half-width of the band (σ) is, within experimental error, constant for the range of concentrations being determined.

In future, because of rapid development in computer techniques the area of the band, i.e. the integrated intensity used in quantitative analysis, will be measured instrumentally. Resolution of overlapping bands will also be done automatically.

REFERENCES

[1] Z. Kęcki, *Fundamentals of Molecular Spectroscopy* (in Polish), 2nd Ed., PWN, Warszawa, 1975.

[2] L. Sobczyk and A. Kisza, *Physical Chemistry for Scientists* (in Polish), PWN, Warszawa, 1975.

[3] G. M. Barrow, *Introduction to Molecular Spectroscopy*, International Student Edition, McGraw-Hill, New York, 1962.

[4] J. R. Dyer, *Application of Absorption Spectroscopy of Organic Compounds*, Prentice-Hall, Englewood Cliffs, New Jersey, 1965.

[5] R. M. Silverstein and G. C. Bassler, *Spectrometric Identification of Organic Compounds*, 2nd Ed., Wiley, New York, 1967.

[6] N. L. Alpert, W. E. Keiser and H. A. Szymanski, *Theory and Practice of Infrared Spectroscopy*, Plenum Press, New York, 1970.

9

Atlas of spectra

J. Terpiński

In order to obtain Raman spectra of the highest quality (without fluorescence bands), all samples were purified (by distillation, sublimation, crystallization or chromatography) before their spectra were recorded. The purified samples were placed in sealed capillary tubes and their spectra recorded with a PHO Coderg spectrometer, with a 6–8 cm^{-1} band-width. The helium–neon laser line at 632.8 nm was used as the exciting radiation. Since photomultiplier response decreases with increasing wavelength (i.e. with increasing distance from the exciting line) some important portions of spectra with $\Delta\nu > 2000$ cm^{-1} were re-recorded, with an argon-ion laser as the exciting source (the blue-green line at 488 nm). In this way an amplification of the intensity of bands near 3000 cm^{-1} was achieved, resulting in an equalization of instrumental sensitivity over the range 0–4000 cm^{-1}, i.e. in the total range under examination. The infrared spectra were recorded with a Perkin-Elmer 577 grating spectrometer with automatic control of slit-width, giving a constant supply of energy to the detector in the range from 4000 to 400 cm^{-1}. Solid samples (organic and inorganic) were examined in the form of pellets made up with potassium bromide. Liquid samples were examined as films formed between potassium bromide plates, or in cells with spacers of suitable thickness.

The following information is given in the legends: state of aggregation of the sample (L—liquid, C—crystal), colour, type of laser (He–Ne or Ar^+) and the assignment of the more important bands to the characteristic vibrations, which allows the identification of essential elements of the structure of the compounds examined. The upper curve corresponds to the infrared spectrum, the lower curve to the Raman spectrum.

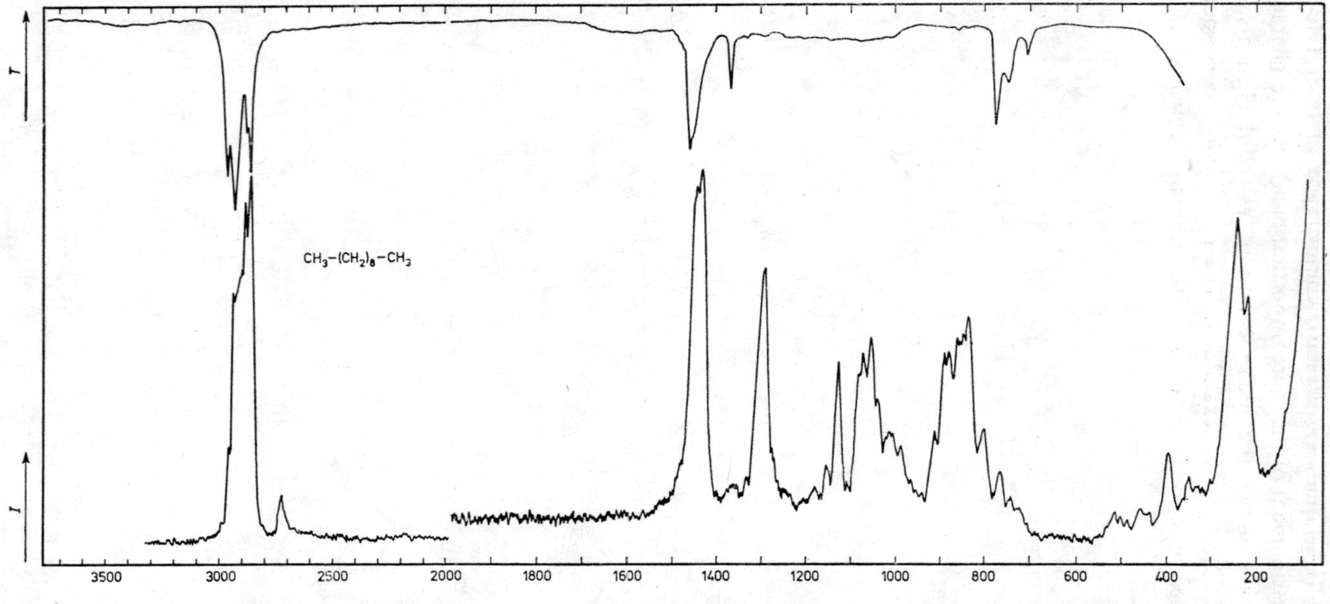

Fig. 9.1—Decane, L (colourless), He–Ne. 2960, ν_{CH_3} as; 2912, ν_{CH_2} as; 2875, ν_{CH_3} sym; 2860, ν_{CH_2} sym; 1470, δ_{CH_3} sym; 1450, 1440, δ_{CH_2}; 1380 (IR), δ_{CH_3} sym; 1305 (R), ω_{CH_2} of numerous neighbouring CH_2 groups; 1080, 1065, 890, 875, 855, 845 (R), skeletal bands characteristic of the linear chain; 720 (IR), ϱ_{CH_2}; 255, 230 (R), chain deformation vibrations.

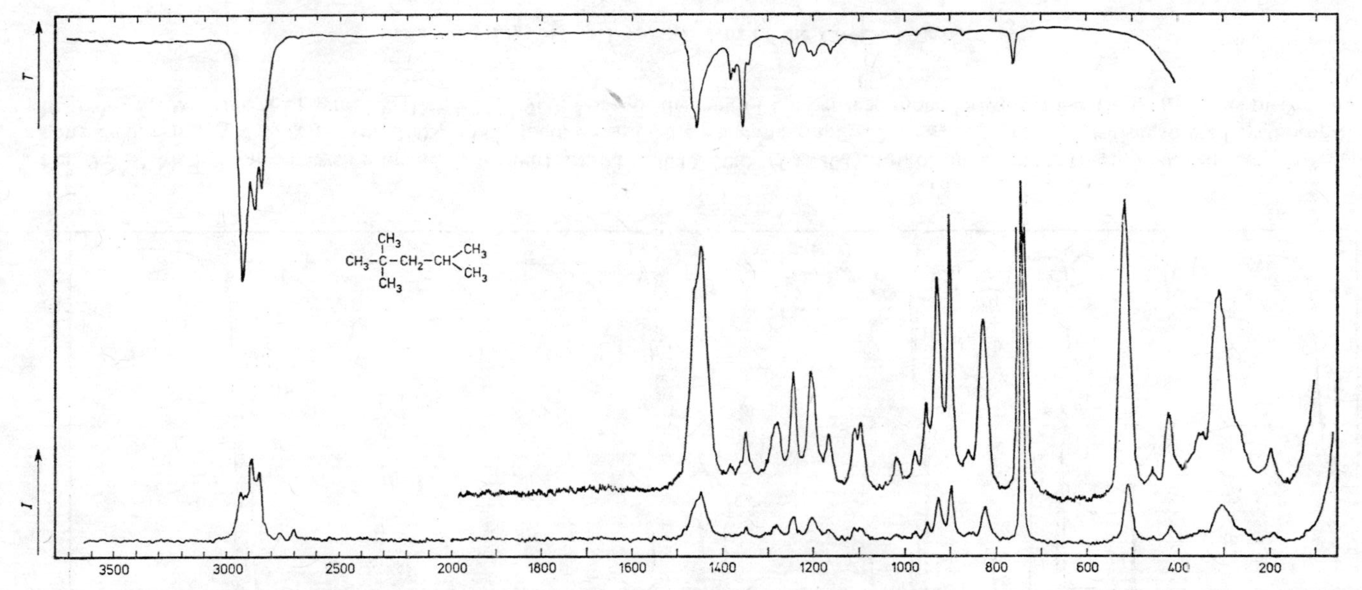

Fig. 9.2—2,2,4-Trimethylpentane (iso-octane) L (colourless), He–Ne. 2960, 2940, ν_{CH_3} as; 2875, ν_{CH_3} sym; 1475, 1470, δ_{CH_3} as; 1390, 1365, δ_{CH_3} sym of the tert-butyl group; 1380, 1360, δ_{CH_3} sym of the isopropyl group; 1245, 1207, $\nu_{C—C}$ of the tert-butyl group; 1170, $\nu_{C—C}$ of the isopropyl group; 955, 925, 900, bands characteristic of the branched chain; 745 (R), $\nu_{skeletal}$ sym of the C—C(C)(C)—C group.

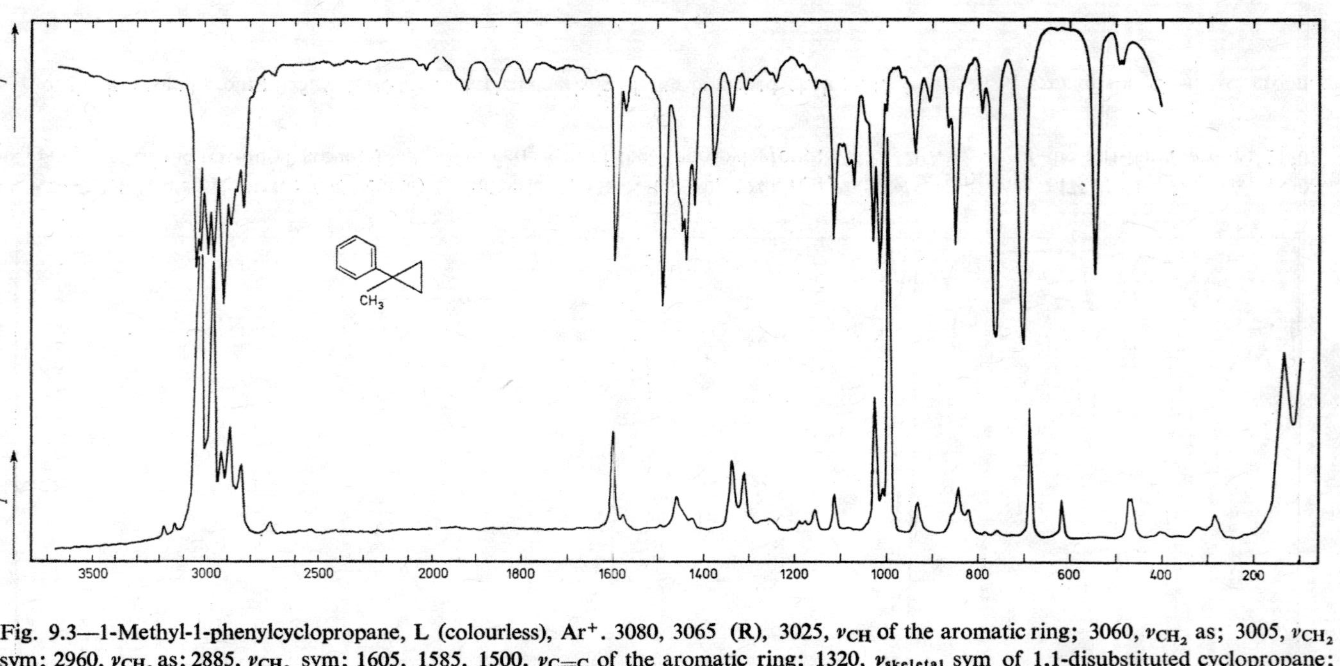

Fig. 9.3—1-Methyl-1-phenylcyclopropane, L (colourless), Ar^+. 3080, 3065 (R), 3025, ν_{CH} of the aromatic ring; 3060, ν_{CH_2} as; 3005, ν_{CH_2} sym; 2960, ν_{CH_3} as; 2885, ν_{CH_3} sym; 1605, 1585, 1500, $\nu_{C=C}$ of the aromatic ring; 1320, $\nu_{skeletal}$ sym of 1,1-disubstituted cyclopropane; 1030, δ_{C-H}, 1000 (R), $\nu_{C=C}$ sym, 755 (IR), γ_{C-H}, 700, δ_{C-C} of the ring of the benzene monoderivative; 845 (IR), $\nu_{skeletal}$ as of the three-membered ring; 685 (R), $\nu_{skeletal}$ sym of the $\mathrm{C{-}\overset{\overset{C}{|}}{\underset{\underset{C}{|}}{C}}{-}C}$ group.

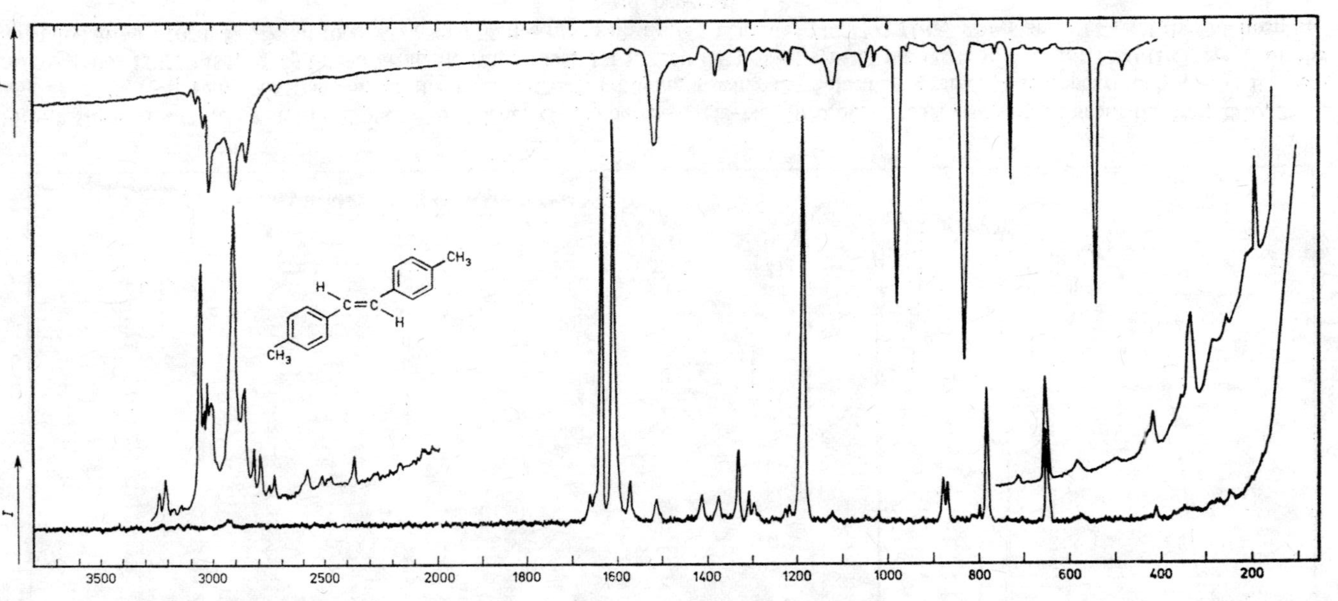

Fig. 9.4—*trans*-4,4′-Dimethylstilbene, L (colourless), He–Ne, Ar⁺. 3070, 3050 3015, ν_{C-H} of the aromatic ring and the olefinic group; 2925, ν_{CH_3} as of the aromatic ring; 2865 ν_{CH_3} sym; 1630, $\nu_{C=C}$ of the conjugated alkene (owing to molecular symmetry, this band is absent from the IR spectrum); 1605, 1565, 1510, 1505, $\nu_{C=C}$ of the aromatic ring; 1327 $\delta_{=C-H}$; 1185 (R), $\nu_{skeletal}$ of the substituent; 970 (IR), γ_{C-H} of the *trans*-disubstituted alkene; 825 (IR), γ_{C-H} of the *p*-substituted benzene.

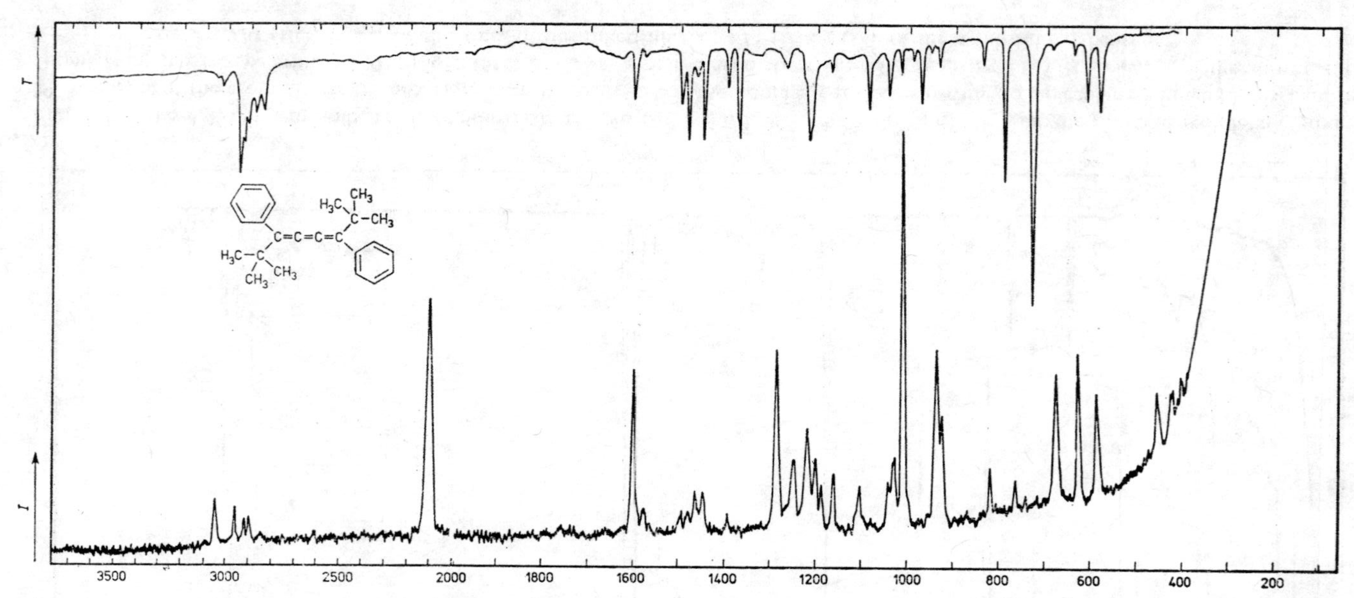

Fig. 9.5-*trans*-1,4-Di-tert-butyl-1,4-diphenylbutatriene, C (colourless), He–Ne. 3070, 3050, 3035, ν_{C-H} of the aromatic ring; 2970, 2960, ν_{CH_3} as; 2860, ν_{CH_3} sym; 2055, ν sym of the C=C=C=C grouping (owing to molecular symmetry, it appears only in the R spectrum); 1600, 1595, 1485, $\nu_{C=C}$ of the aromatic ring; 1395, 1365, 1360 (IR), δ_{CH_3} sym of the tert-butyl group; 1200 (IR), ν_{C-C} of the tert-butyl group; 1020, δ_{C-H} and 1000, $\nu_{C=C}$ sym of the monosubstituted benzene; 770, 705 (IR), γ_{C-H} and δ_{C-C} of the monosubstituted benzene.

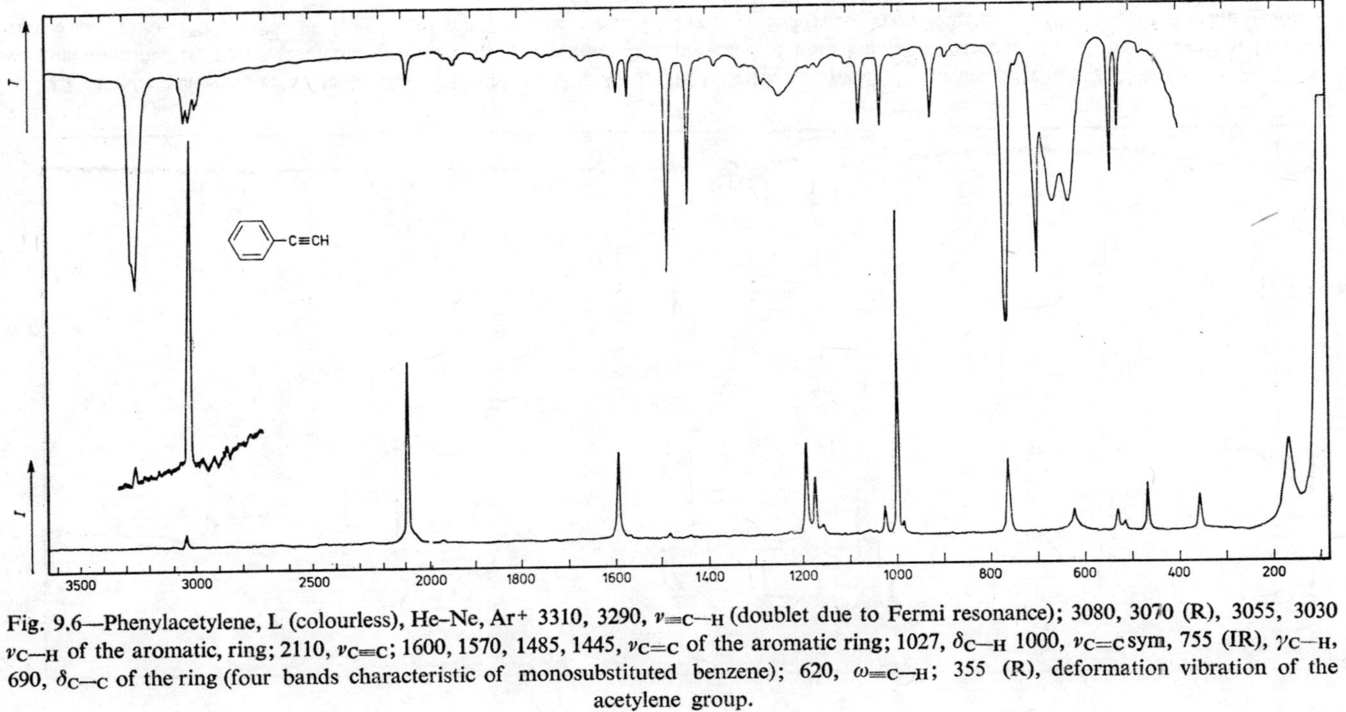

Fig. 9.6—Phenylacetylene, L (colourless), He–Ne, Ar^{+} 3310, 3290, $\nu_{\equiv C-H}$ (doublet due to Fermi resonance); 3080, 3070 (R), 3055, 3030 ν_{C-H} of the aromatic, ring; 2110, $\nu_{C\equiv C}$; 1600, 1570, 1485, 1445, $\nu_{C=C}$ of the aromatic ring; 1027, δ_{C-H} 1000, $\nu_{C=C}$ sym, 755 (IR), γ_{C-H}, 690, δ_{C-C} of the ring (four bands characteristic of monosubstituted benzene); 620, $\omega_{\equiv C-H}$; 355 (R), deformation vibration of the acetylene group.

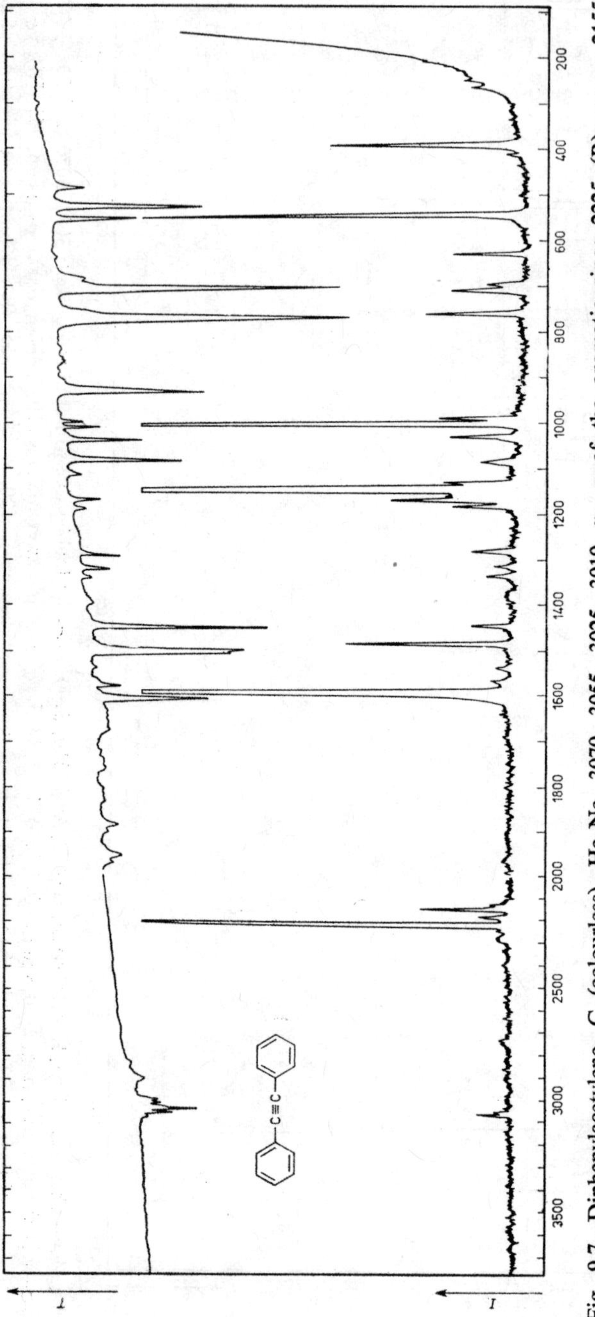

Fig. 9.7—Diphenylacetylene, C (colourless), He–Ne. 3070, 3055, 3025, 3010, ν_{C-H} of the aromatic ring; 2225 (R), $\nu_{C\equiv C}$, 2155, overtone amplified by Fermi resonance with the $\nu_{C\equiv C}$ vibration (the presence of both bands is characteristic of the disubstituted acetylene); 1600, 1590, 1500, 1490, 1445, $\nu_{C=C}$ of the aromatic ring; 1025, δ_{C-H}, 995, $\nu_{C=C}$ sym, 755, γ_{C-H}, 690, δ_{C-C} of the ring (monosubstituted benzene)

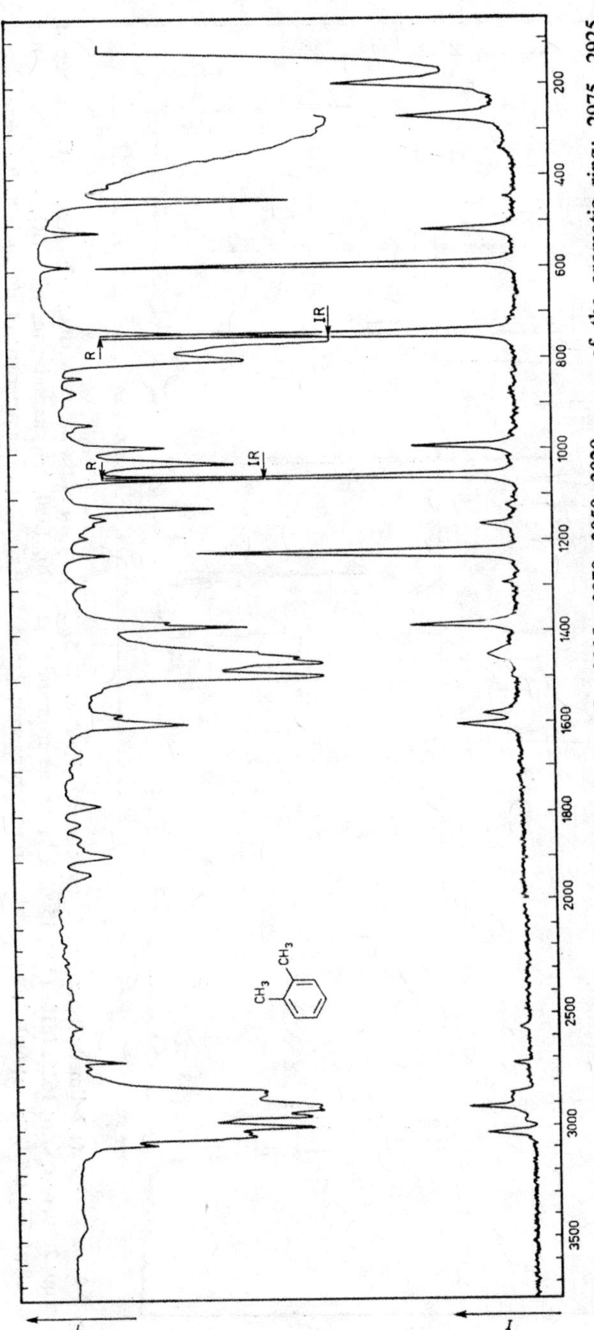

Fig. 9.8—*o*-Xylene (1,2-dimethylbenzene), L (colourless), He–Ne. 3085, 3070, 3050, 3020, $\nu_{C—H}$ of the aromatic ring; 2975, 2925, ν_{CH_3} at the ring; 2880, 2860, ν_{CH_3} sym (the increased number of bands in this region is due to vibration coupling of neighbouring methyl groups); 1610, 1580, 1497, $\nu_{C=C}$ of the aromatic ring; 1470, 1455, 1385, δ_{CH_3}; 1050 (R), $\delta_{C—H}$, 740 (IR), $\gamma_{C—H}$ *o*-disubstituted benzene; 1225, 735, 580 (R), $\nu_{skeletal}$ characteristic of the *o*-dialkylbenzene.

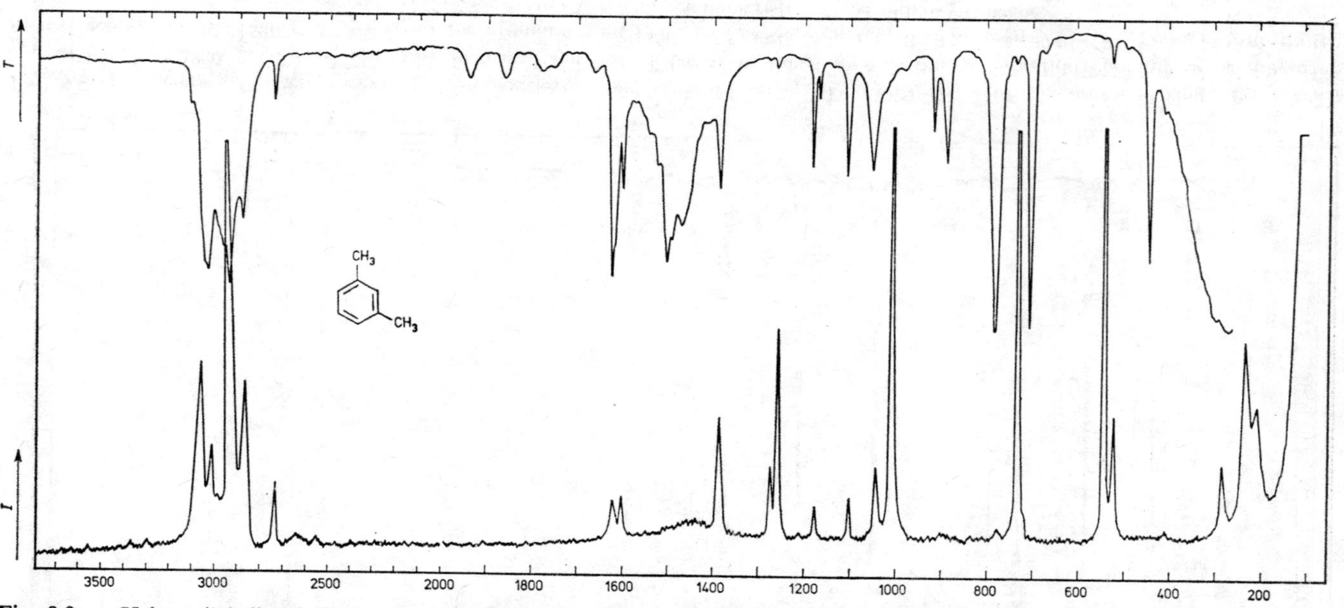

Fig. 9.9—*m*-Xylene (1,3-dimethylbenzene), L (colourless), He–Ne. 3055, 3020, ν_{C-H} of the aromatic ring; 2920, ν_{CH_3} as (CH_3 group at the ring); 2870, ν_{CH_3} sym; 1625, 1610, 1590, 1500, $\nu_{C=C}$ of the aromatic ring; 1470, 1380, δ_{CH_3} as and sym; 1000 (R), $\nu_{C=C}$ sym of the ring, 770 (IR), γ_{C-H}; 690, δ_{C-C} (*m*-disubstituted benzene); 725, 535 (R), $\nu_{skeletal}$ characteristic of the *m*-dialkylbenzene.

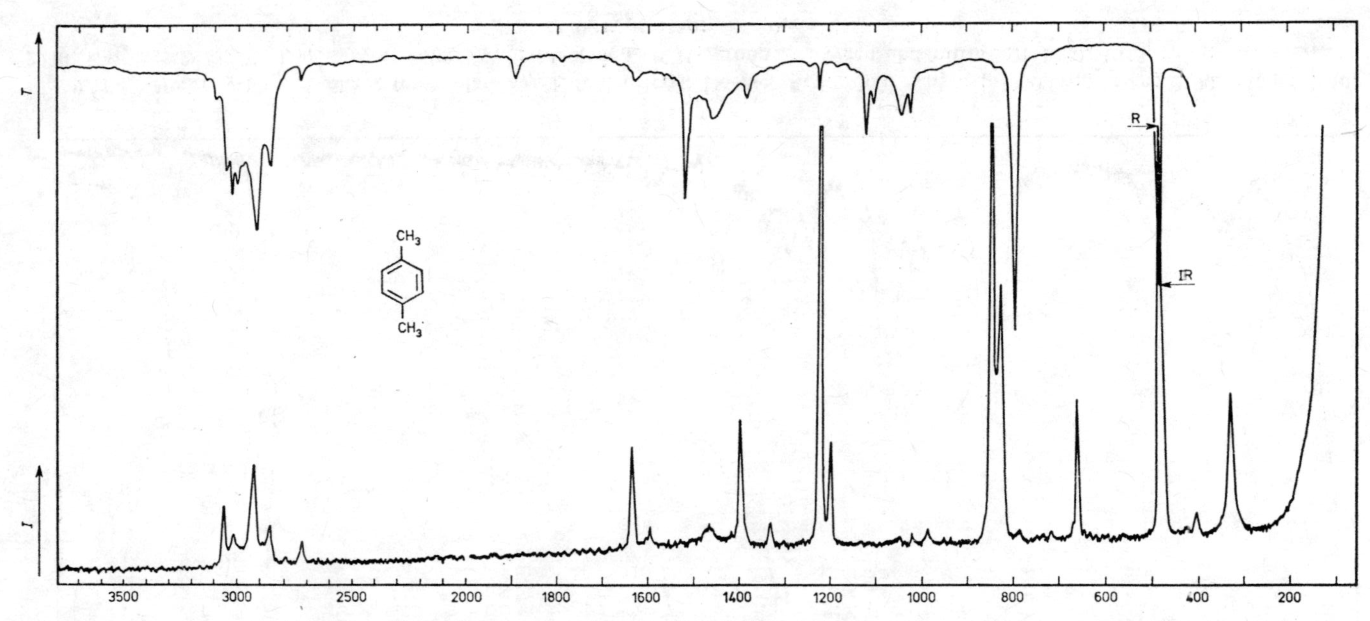

Fig. 9.10—*p*-Xylene (1,4-dimethylbenzene), L (colourless), He–Ne. 3065, 3060, 3020, 3015, $\nu_{C—H}$ of the aromatic ring; 2920, ν_{CH_3} as (CH_3 group at the ring); 2865, ν_{CH_3} sym; 1620, 1515, $\nu_{C=C}$ of the aromatic ring; 1450, 1375, δ_{CH_3} as and sym; 825 $\nu_{skeletal}$; 795, $\gamma_{C—H}$ of the *p*-disubstituted benzene; 1205, $\nu_{skeletal}$ characteristic of the *p*-dialkylbenzene.

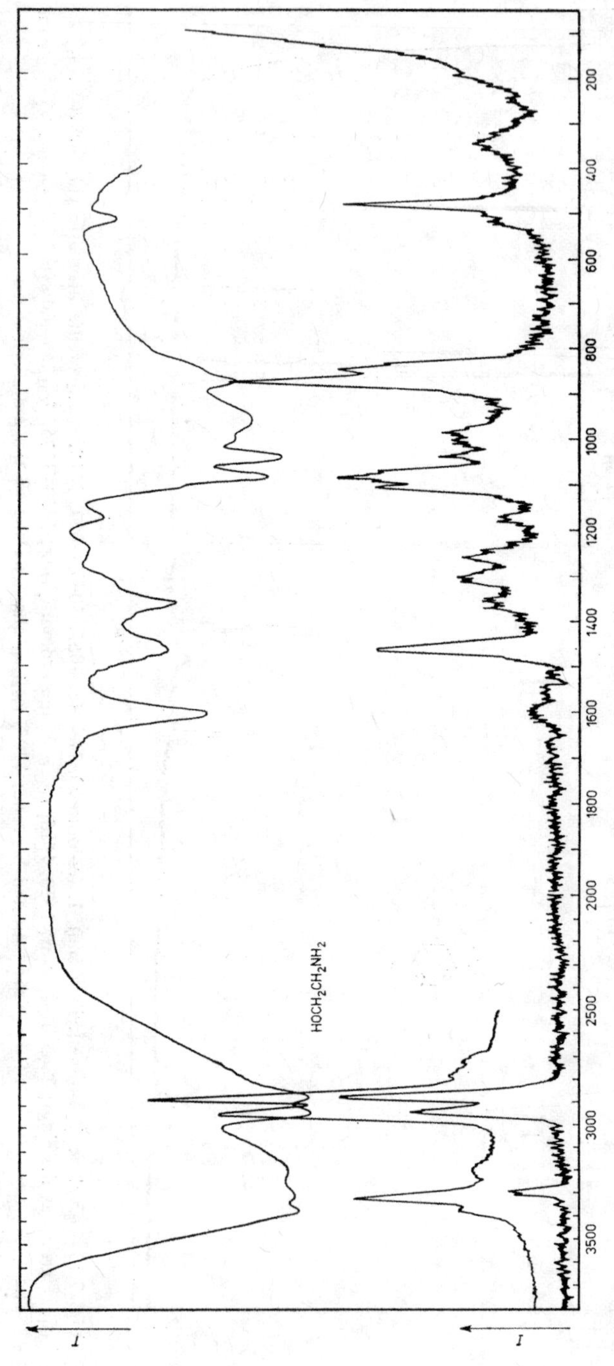

Fig. 9.11—2-Hydroxy-1-aminoethane (ethanolamine), L (colourless), He–Ne, Ar^+. 3360, ν_{NH_2} as; 3300, ν_{NH_2} sym; 3190, ν_{O-H}; 2940, 2930, ν_{CH_3} as; 2875, 2860, ν_{CH_2} sym; 1600, δ_{NH_2}; 1460, δ_{CH_2} (CH_2 group at oxygen and nitrogen); 1080, 1035, ν_{C-N} and ν_{CCO} as; 870, ν_{CCO} sym; 480, δ_{CCO}.

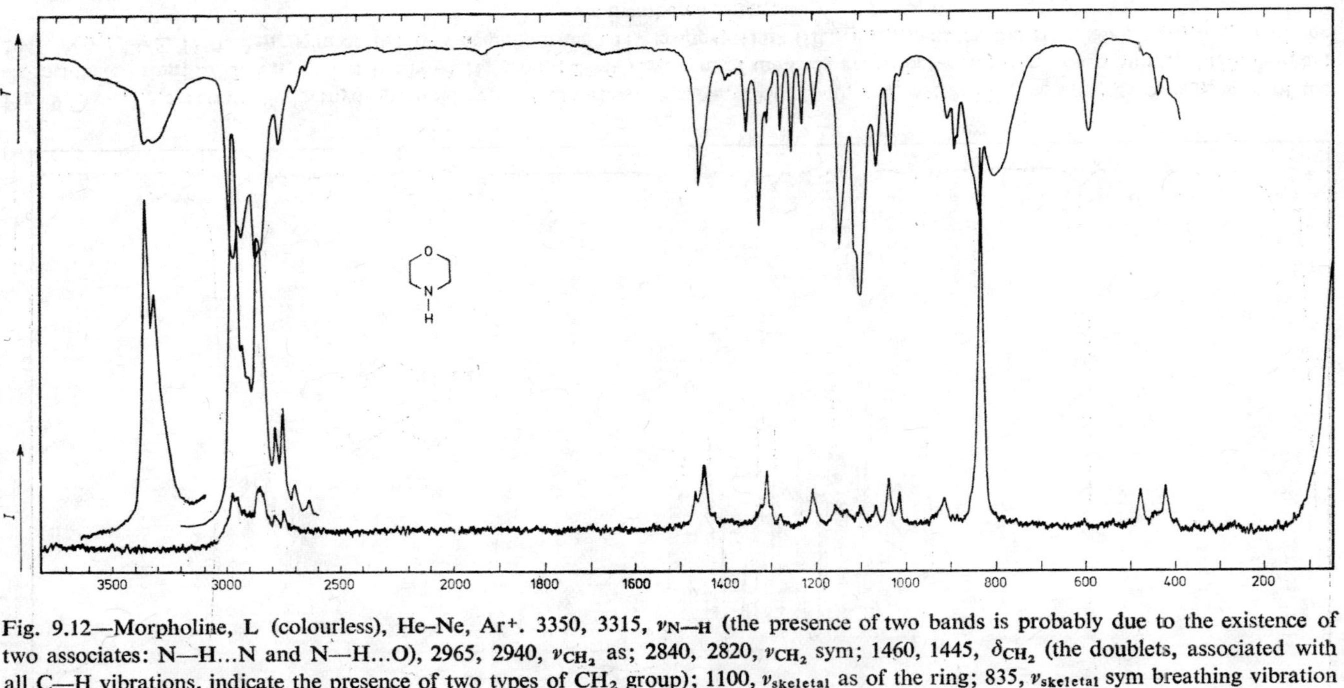

Fig. 9.12—Morpholine, L (colourless), He–Ne, Ar+. 3350, 3315, $\nu_{N—H}$ (the presence of two bands is probably due to the existence of two associates: N—H...N and N—H...O), 2965, 2940, ν_{CH_2} as; 2840, 2820, ν_{CH_2} sym; 1460, 1445, δ_{CH_2} (the doublets, associated with all C—H vibrations, indicate the presence of two types of CH_2 group); 1100, $\nu_{skeletal}$ as of the ring; 835, $\nu_{skeletal}$ sym breathing vibration of the six-membered ring.

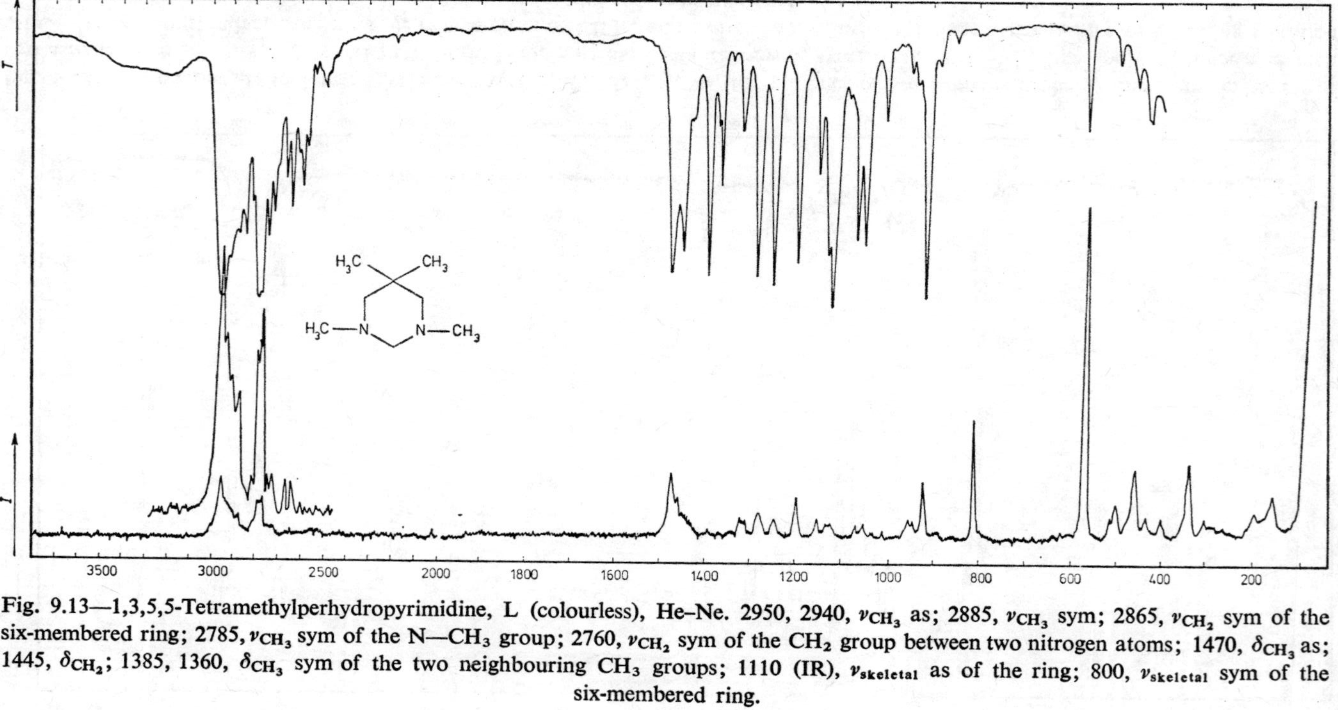

Fig. 9.13—1,3,5,5-Tetramethylperhydropyrimidine, L (colourless), He–Ne. 2950, 2940, ν_{CH_3} as; 2885, ν_{CH_3} sym; 2865, ν_{CH_2} sym of the six-membered ring; 2785, ν_{CH_3} sym of the N—CH_3 group; 2760, ν_{CH_2} sym of the CH_2 group between two nitrogen atoms; 1470, δ_{CH_3} as; 1445, δ_{CH_2}; 1385, 1360, δ_{CH_3} sym of the two neighbouring CH_3 groups; 1110 (IR), $\nu_{skeletal}$ as of the ring; 800, $\nu_{skeletal}$ sym of the six-membered ring.

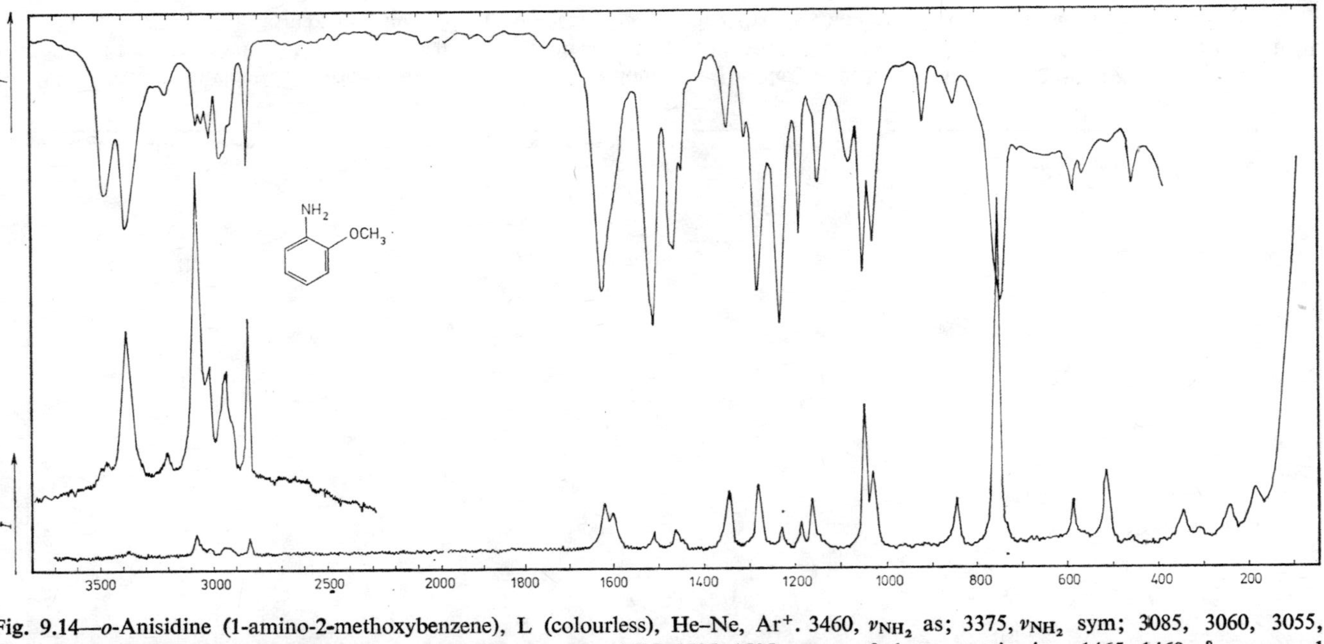

Fig. 9.14—*o*-Anisidine (1-amino-2-methoxybenzene), L (colourless), He–Ne, Ar^+. 3460, ν_{NH_2} as; 3375, ν_{NH_2} sym; 3085, 3060, 3055, 3020, ν_{C-H} of the aromatic ring; 2945, ν_{CH_3} as; 2840, ν_{CH_3} sym; 1615, 1595, 1505, $\nu_{C=C}$ of the aromatic ring; 1465, 1460, δ_{CH_3} as and sym of the OCH_3 group; 1275, ν_{C-N} of the aromatic amine; 1225, ν_{C-O} of the aryl ether; 1045 (R), δ_{C-H}; 740 (IR), γ_{C-H} (*o*-disubstituted benzene).

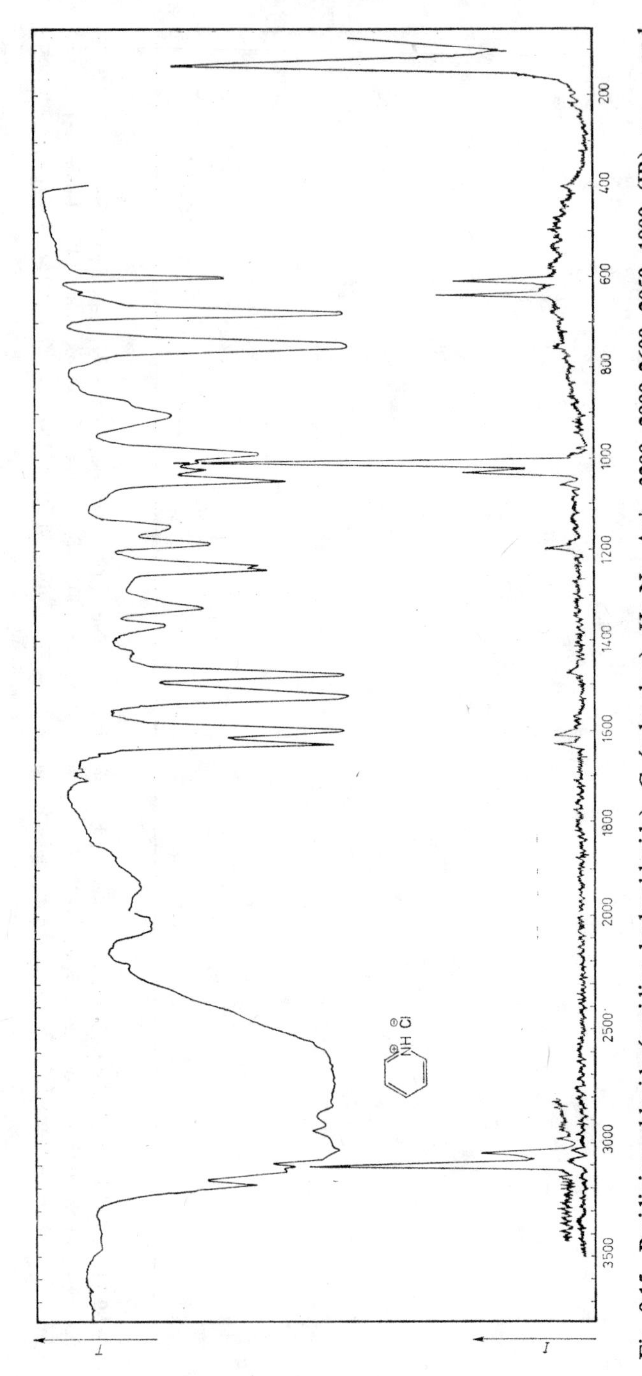

Fig. 9.15—Pyridinium chloride (pyridine hydrochloride), C (colourless), He–Ne, Ar^+. 3200, 2900–2600, 2050, 1900 (IR), $\nu_{N—H}$ and overtones of the amine salt; 3090, 3030, $\nu_{C—H}$ of the aromatic ring; 1630, 1600, 1525, 1480, $\nu_{skeletal}$ of the heteroaromatic ring; 1030, $\delta_{C—H}$; 1010, $\nu_{skeletal}$ sym of the ring; 755, $\gamma_{C—H}$; 685, δ of the ring (bands characteristic of the unsubstituted pyridine).

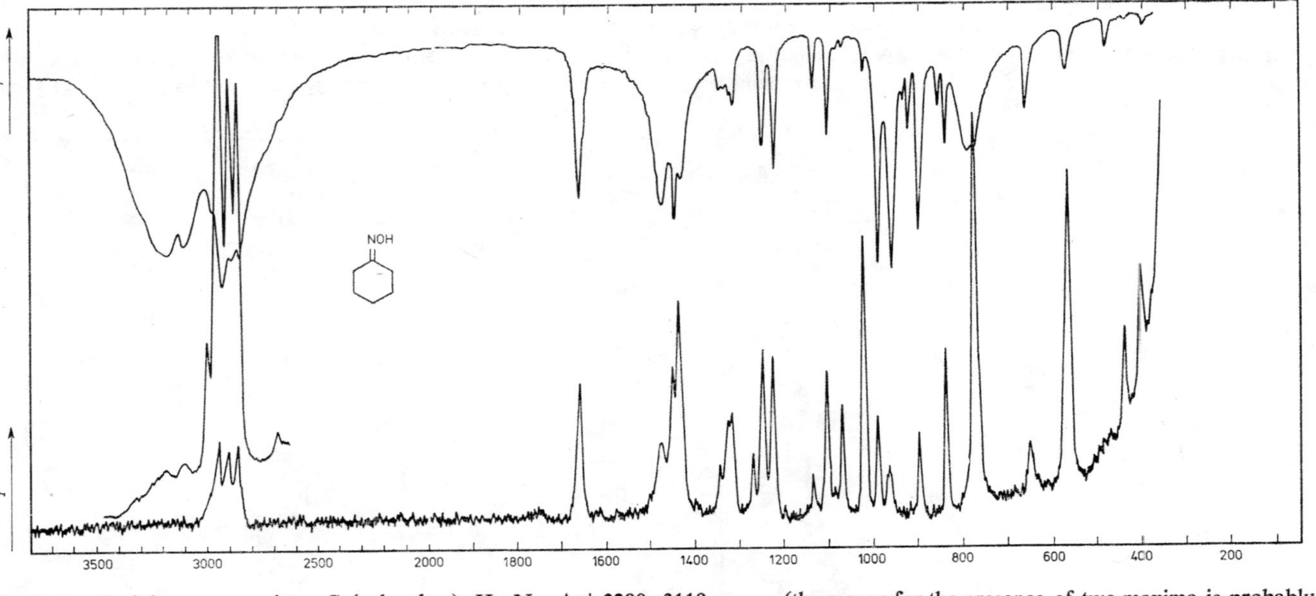

Fig. 9.16—Cyclohexanone oxime, C (colourless), He–Ne, Ar^+ 3200, 3110, ν_{O-H} (the reason for the presence of two maxima is probably dimerization of the oxime); 2950, 2900, ν_{CH_2} as, 2860, ν_{CH_2} sym, 1450, 1435, δ_{CH_2} of the six-membered ring; 1660, $\nu_{C=N}$ (the higher frequency and the intensity of the band in both spectra prove that this band is not the $\nu_{C=C}$ band); 960, $\nu_{N=O}$; 775, $\nu_{skeletal}$ sym (breathing) of the six-membered ring (the presence of the C=N grouping in the molecule shifts the band characteristic of cyclohexane from its normal position).

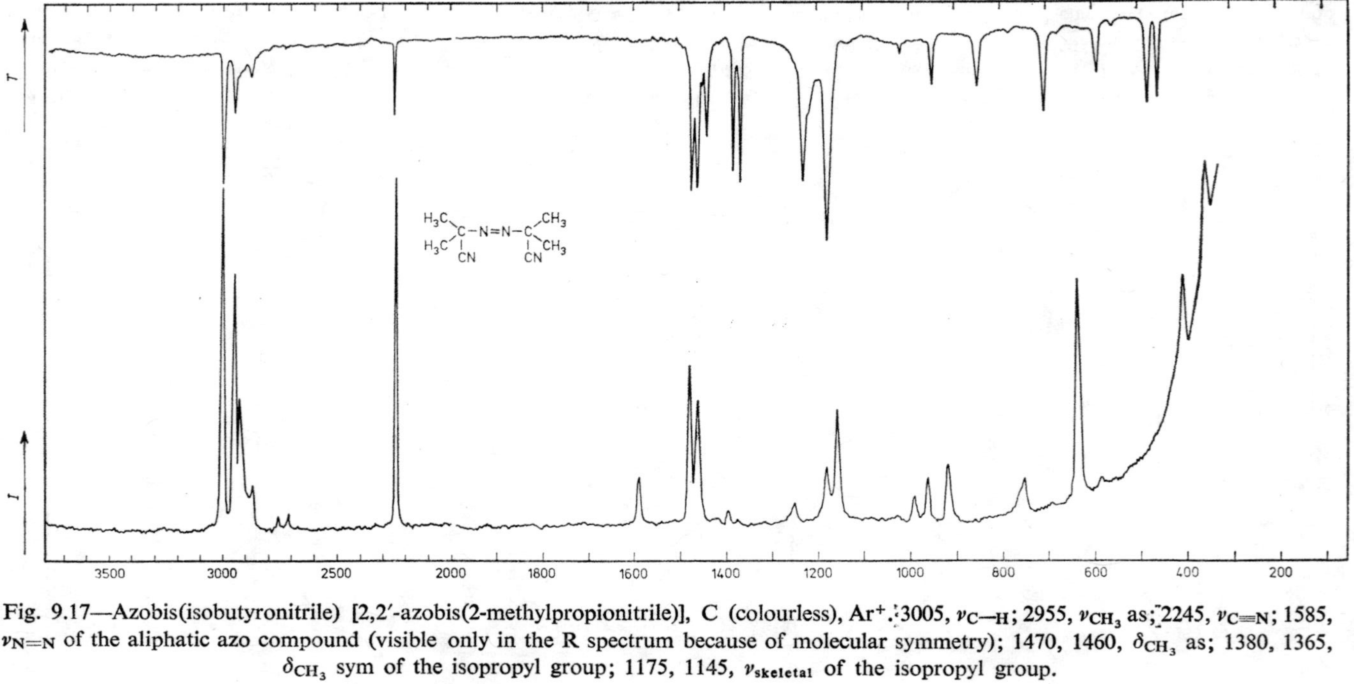

Fig. 9.17—Azobis(isobutyronitrile) [2,2′-azobis(2-methylpropionitrile)], C (colourless), Ar⁺.: 3005, ν_{C-H}; 2955, ν_{CH_3} as; 2245, $\nu_{C\equiv N}$; 1585, $\nu_{N=N}$ of the aliphatic azo compound (visible only in the R spectrum because of molecular symmetry); 1470, 1460, δ_{CH_3} as; 1380, 1365, δ_{CH_3} sym of the isopropyl group; 1175, 1145, $\nu_{skeletal}$ of the isopropyl group.

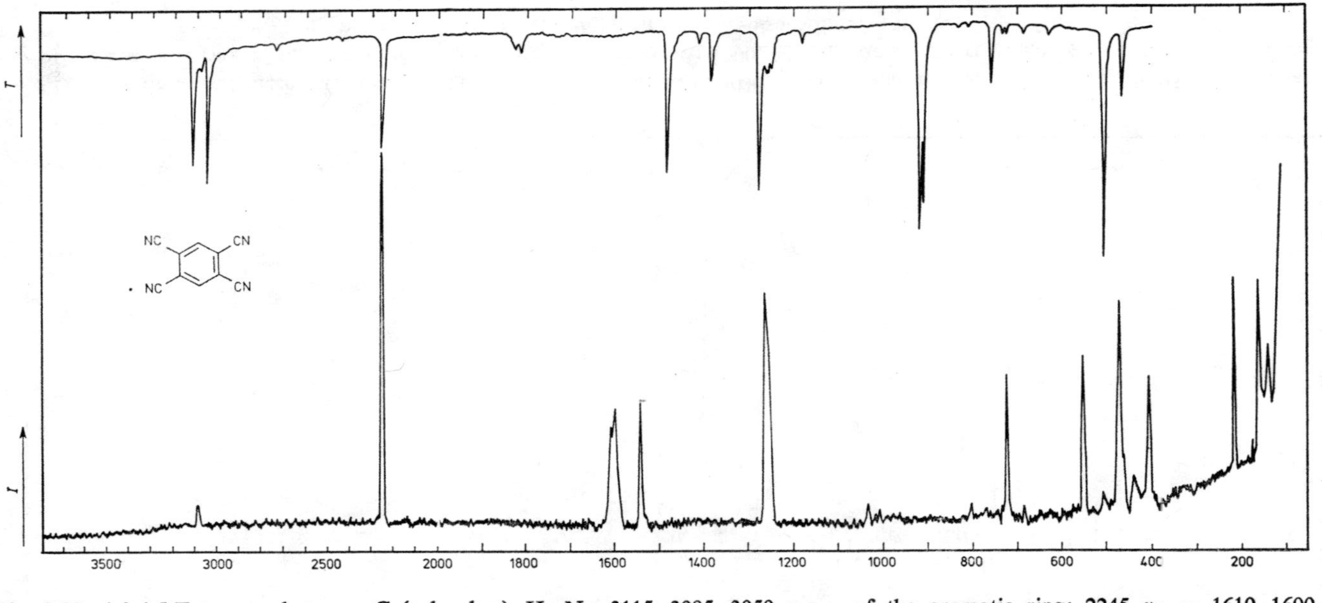

Fig. 9.18—1,2,4,5-Tetracyanobenzene, **C** (colourless), He–Ne. 3115, 3085, 3050, $\nu_{C—H}$ of the aromatic ring; 2245, $\nu_{C\equiv N}$, 1610, 1600, 1545, 1485, $\nu_{C\equiv C}$ of the aromatic ring; 920, 910, $\gamma_{C—H}$ isolated hydrogen atoms at the ring. The band position does not give an exact identification of the substitution type. The high symmetry of the molecule is evident (band positions in the IR spectrum differ from those in the **R** spectrum). The bands corresponding to the combination tones (1830 and 1815) suggest the presence of the 1,2,4,5-tetra-substituted benzene (cf. Fig. 4.5, p. 106).

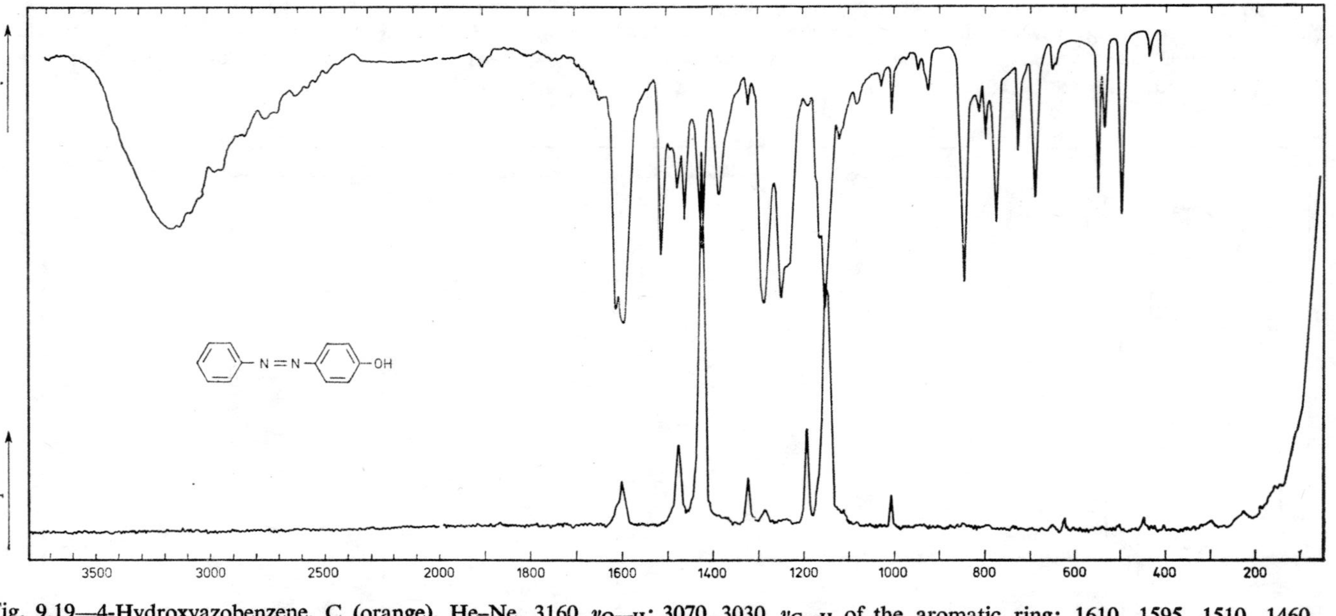

Fig. 9.19—4-Hydroxyazobenzene, C (orange), He–Ne. 3160, ν_{O-H}; 3070, 3030, ν_{C-H} of the aromatic ring; 1610, 1595, 1510, 1460, $\nu_{C=C}$ of the aromatic rings; 1420, $\nu_{N=N}$ of the aromatic azo compound; 1000, $\nu_{skeletal}$ sym of the ring, 770, γ_{C-H}; 685, δ_{C-C} of the monosubstituted ring; 840, γ_{C-H} of the *p*-disubstituted benzene.

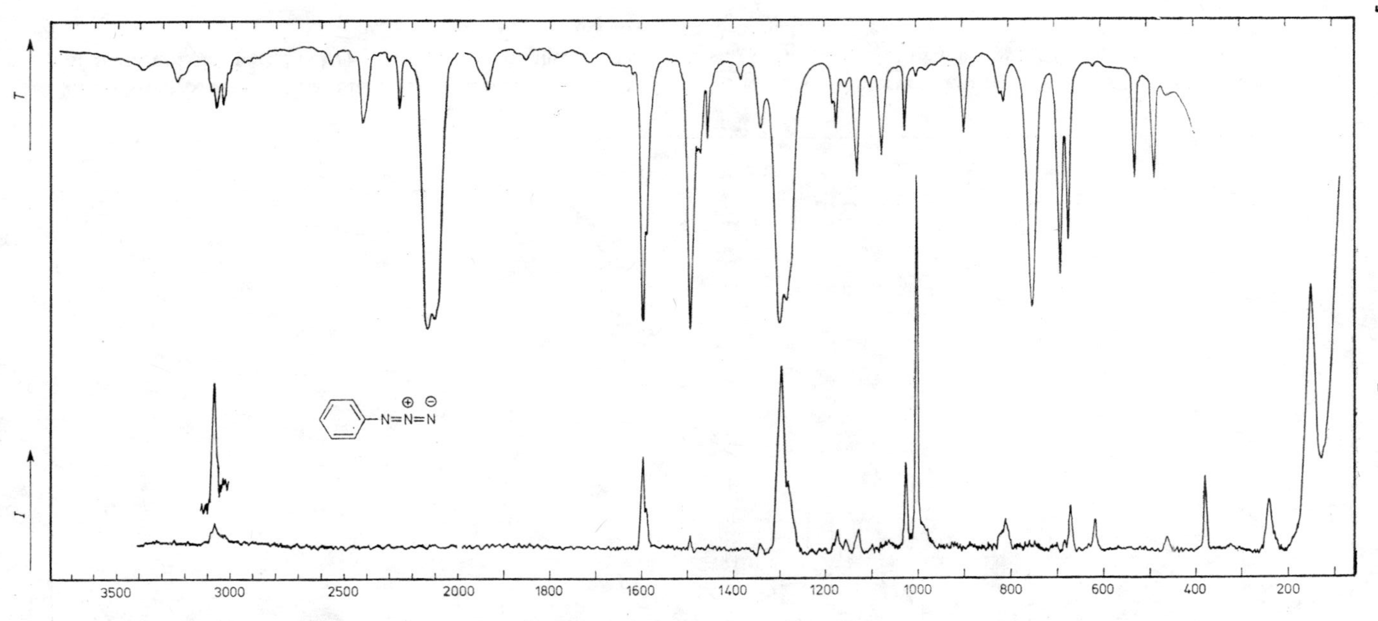

Fig. 9.20—Phenyl azide, L (yellow), He–Ne. 3080, 3070, 3060, 3025, ν_{C-H} of the aromatic ring; 2145, 2100, $\nu_{N=N-N}$ as (splitting due to Fermi resonance); 1595, 1585, 1545, $\nu_{C=C}$ of the aromatic ring) 1295, $\nu_{N=N-N}$ sym; 1025, δ_{C-H}; 1000, $\nu_{C=C}$ sym of the ring, 750, γ_{C-H}; 685, δ_{C-C} of the ring (monosubstituted benzene).

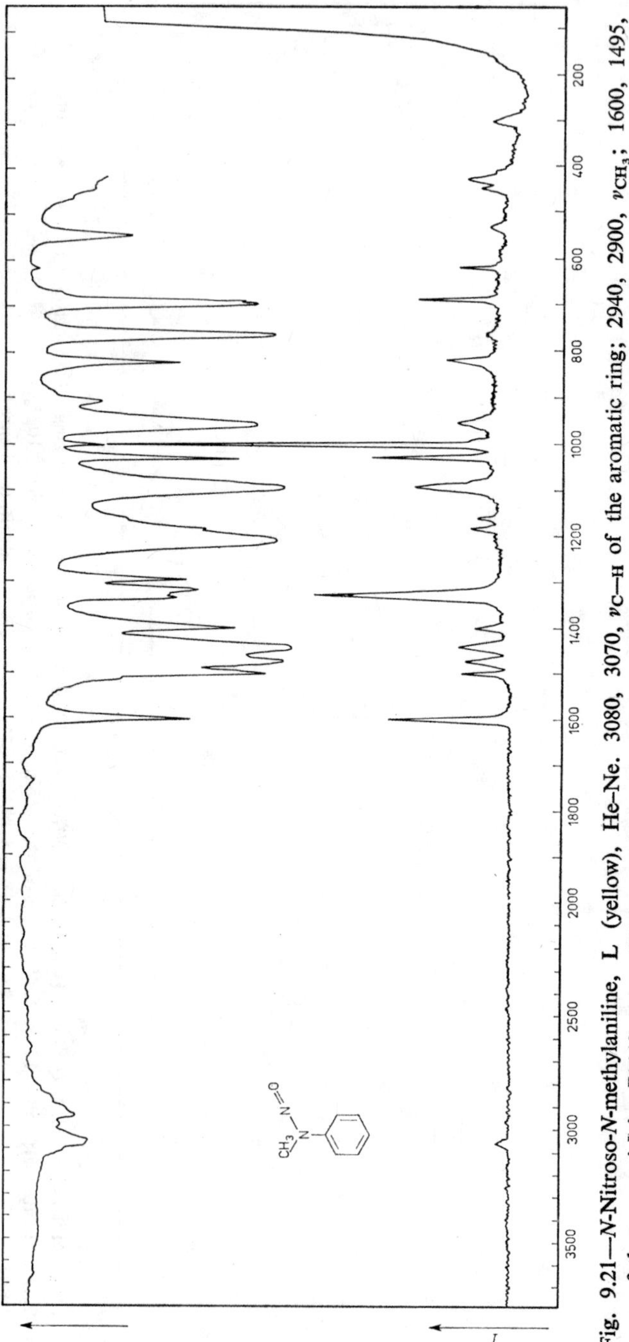

Fig. 9.21—*N*-Nitroso-*N*-methylaniline, **L** (yellow), He–Ne. 3080, 3070, ν_{C-H} of the aromatic ring; 2940, 2900, ν_{CH_3}; 1600, 1495, $\nu_{C=C}$ of the aromatic ring; 1440, $\nu_{N=O}$; 1325, ν_{CN} of the aromatic amine; 1025, δ_{C-H} sym of the ring; 760, γ_{C-H}; 690, δ_{C-C} (of the monosubstituted benzene).

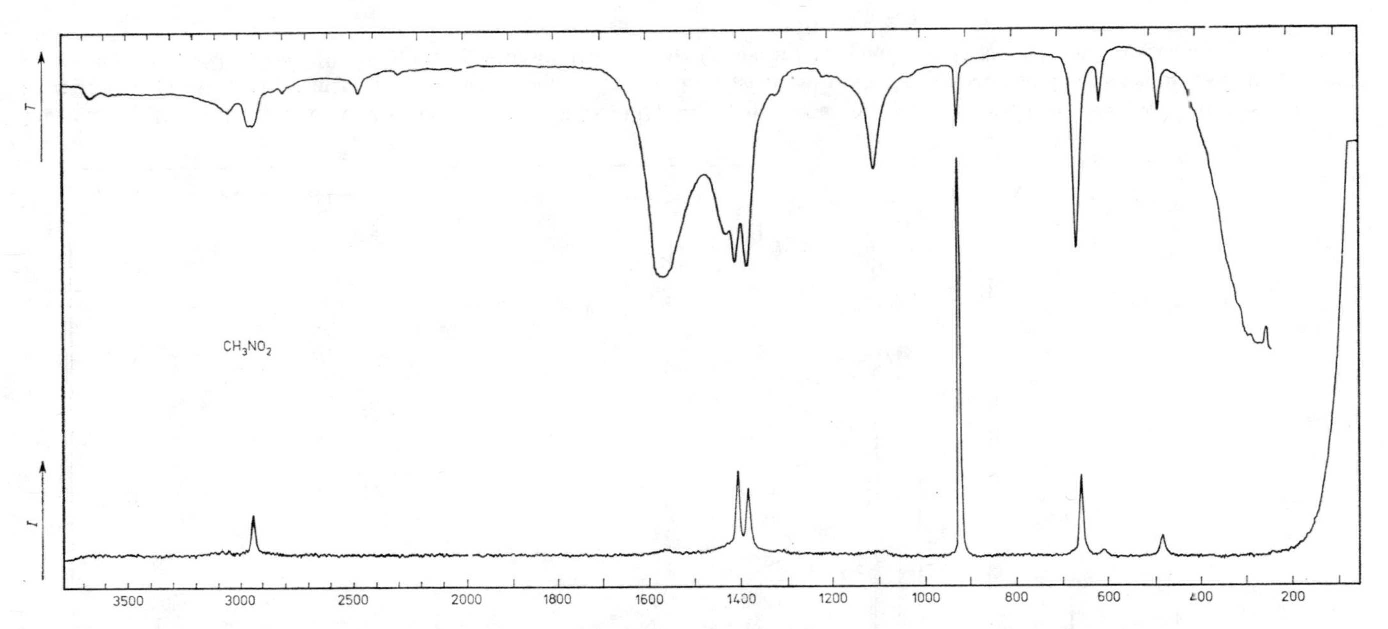

Fig. 9.22—Nitromethane, L (yellow), He–Ne. 2980, ν_{CH_3} as; 2965, ν_{CH_3} sym (the rise in frequency is caused by the NO_2 group); 1570, ν_{NO_2} as, of the aliphatic nitro group; 1425, δ_{CH_3} as; 1400, 1380, doublet due to the coupling of the δ_{CH_3} vibrations with the ν_{NO_2} vibrations; 920, $\nu_{C—N}$; 655, δ_{NO_2}; 605, γ_{NO_2}; 480, ϱ_{NO_2}

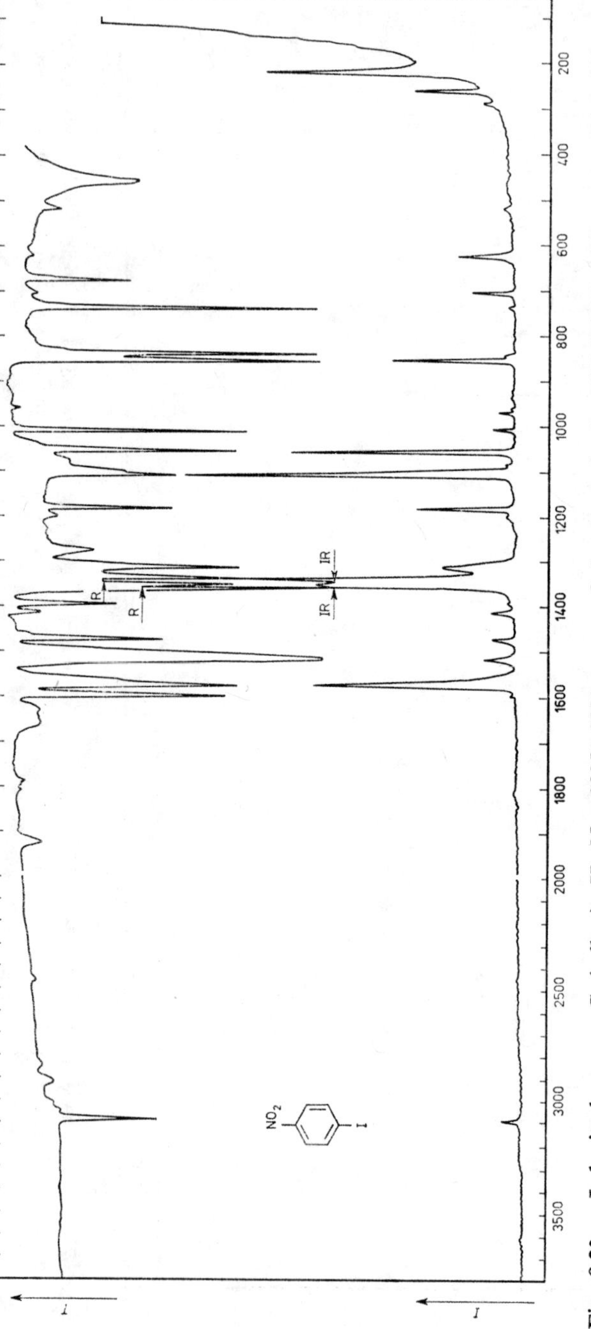

Fig. 9.23—*p*-Iodonitrobenzene, C (yellow), He–Ne. 3085, 3030, $\nu_{C—H}$ of the aromatic ring; 1595, 1570, 1470, $\nu_{C=C}$ of the aromatic ring; 1510, ν_{NO_2} as, of the aromatic nitro group; 1345, ν_{NO_2} sym, the high intensity in the R spectrum is characteristic of an aromatic nitro compound; 860, $\nu_{C—N}$; 850, $\gamma_{C—H}$ of the *p*-substituted benzene (confirmed by bands at 1900 and 1800 associated with combination tones); 1055, $\nu_{skeletal}$ of the *p*-derivative of iodobenzene.

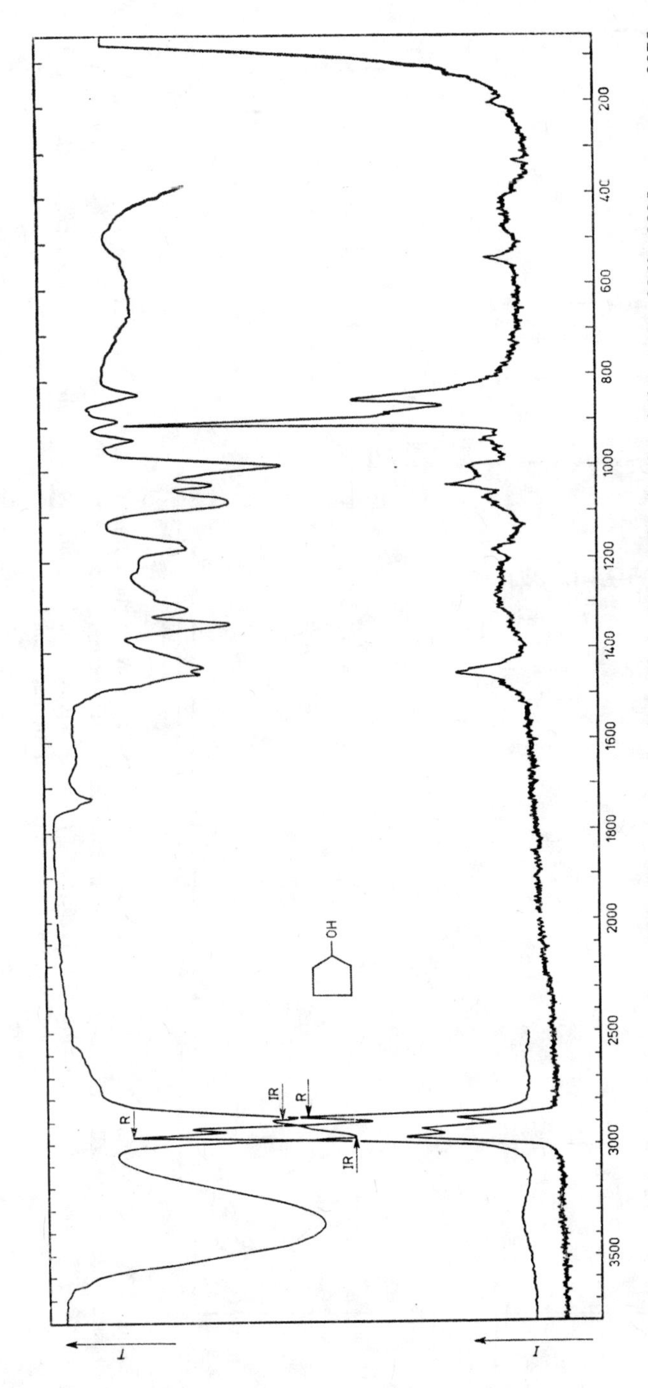

Fig. 9.24—Cyclopentanol, L (colourless), He–Ne, Ar^+. 3350, ν_{O-H} of the intramolecular bonded group; 2960, 2925, ν_{CH_2} as; 2875, ν_{CH_2} sym; 1450, δ_{CH_2}; 1070, ν_{C-O} of the equatorial OH group; 1030, ν_{C-O} of the axial OH group; 955, ν as and 895, ν sym of the five-membered ring.

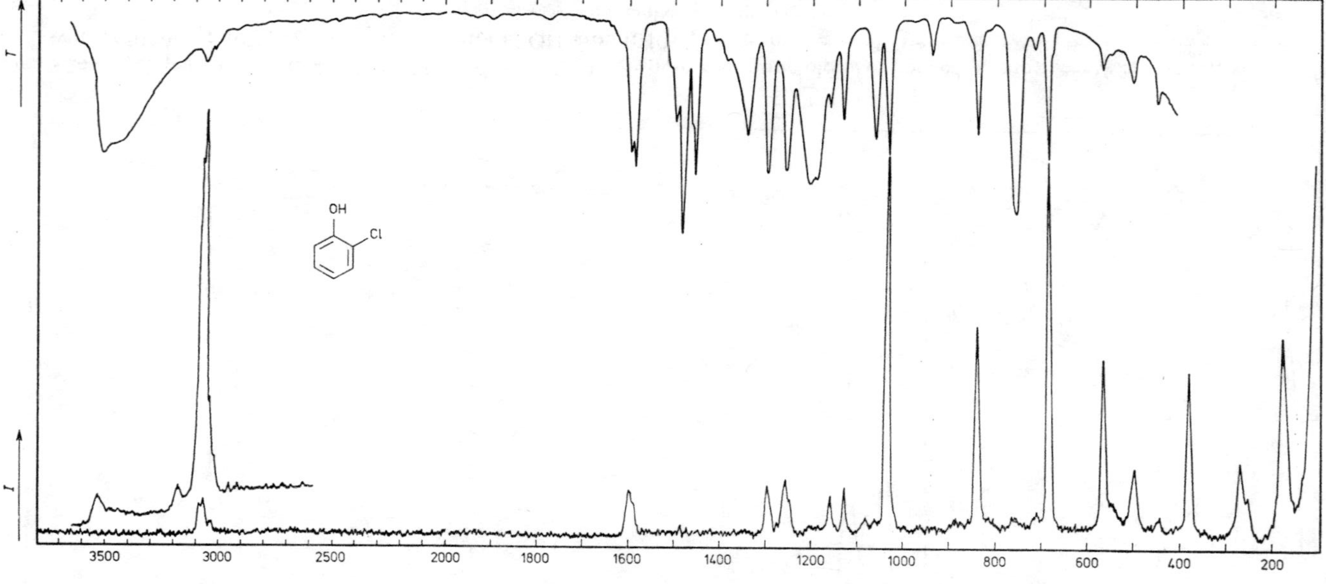

Fig. 9.25—*o*-Chlorophenol, L (colourless), He–Ne, Ar^{+}. 3510, ν_{O-H} of the intramolecular bonded OH group; 3440, ν_{O-H} of the intramolecular bonded OH group (practically invisible in the R spectrum); 3090, 3080, 3050, ν_{C-H} of the aromatic ring; 1600, 1585, 1490, 1480, $\nu_{C=C}$ of the aromatic ring; 1195, 1185, ν_{C-O} of a phenol, 1055, $\nu_{skeletal}$ of the *o*-substituted benzene; 1030, δ_{C-H} of the *o*-substituted benzene.

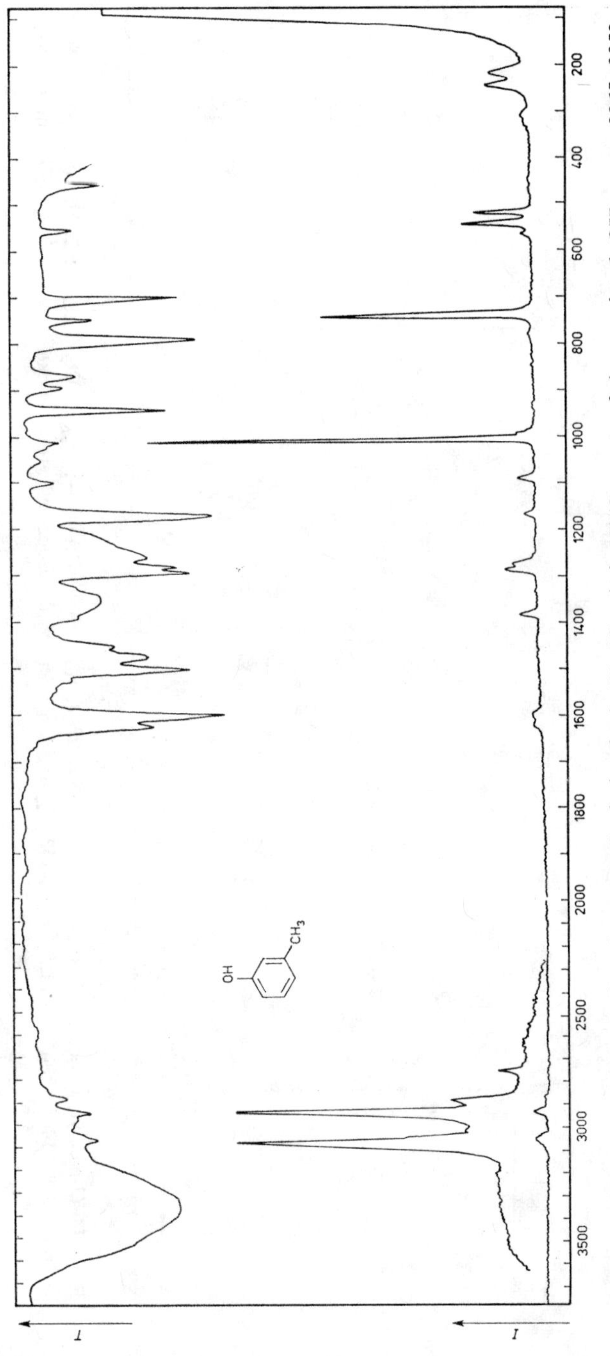

Fig. 9.26—*m*-Cresol (1-methyl-3-hydroxybenzene), L (colourless), He–Ne, Ar^+. 3340, ν_{O-H} of the associated OH group; 3065, 3050, ν_{C-H} of the aromatic ring; 2925, ν_{CH_3} as, of the CH_3 group at the aromatic ring; 2865, ν_{CH_3} sym; 1615, 1595, 1590, 1490, $\nu_{C=C}$ of the aromatic ring; 1160, ν_{C-O} of a phenol; 1000, $\nu_{C=C}$ sym of the ring; 775, γ_{C-H} of the three neighbouring hydrogen atoms; 685, δ_{C-C} of the *m*-substituted benzene.

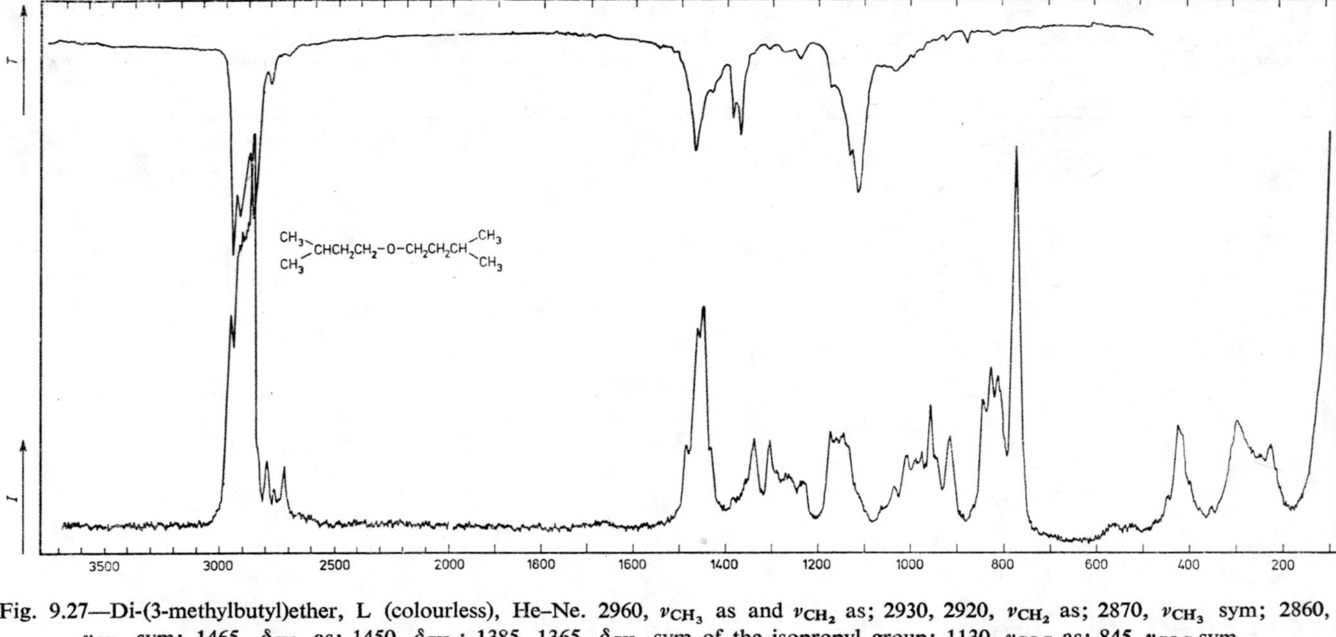

Fig. 9.27—Di-(3-methylbutyl)ether, L (colourless), He–Ne. 2960, ν_{CH_3} as and ν_{CH_2} as; 2930, 2920, ν_{CH_2} as; 2870, ν_{CH_3} sym; 2860, ν_{CH_2} sym; 1465, δ_{CH_3} as; 1450, δ_{CH_2}; 1385, 1365, δ_{CH_3} sym of the isopropyl group; 1130, ν_{COC} as; 845, ν_{COC} sym.

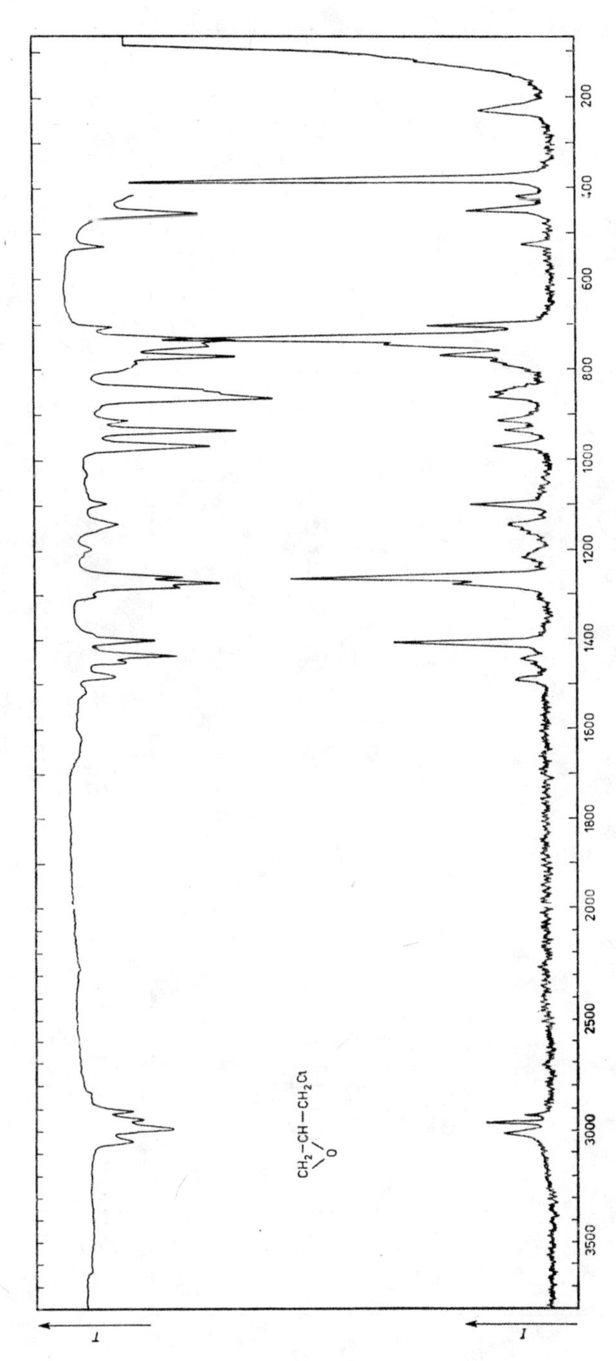

Fig. 9.28—1-Chloro-2,3-epoxypropane, L (colourless), He–Ne. 3060, ν_{CH_2} as, 3010, ν_{CH_2} sym of the three-membered ring; 2965, ν_{CH_2} as, 2930, ν_{CH_2} sym of the CH_2Cl group (the frequency increase is caused by the chlorine atom); 1265, ω_{CH_2} of the CH_2Cl group; 1255, $\nu_{skeletal}$ sym of the three-membered ring; 850 $\nu_{skeletal}$ as of the ring; 725, $\nu_{C—Cl}$.

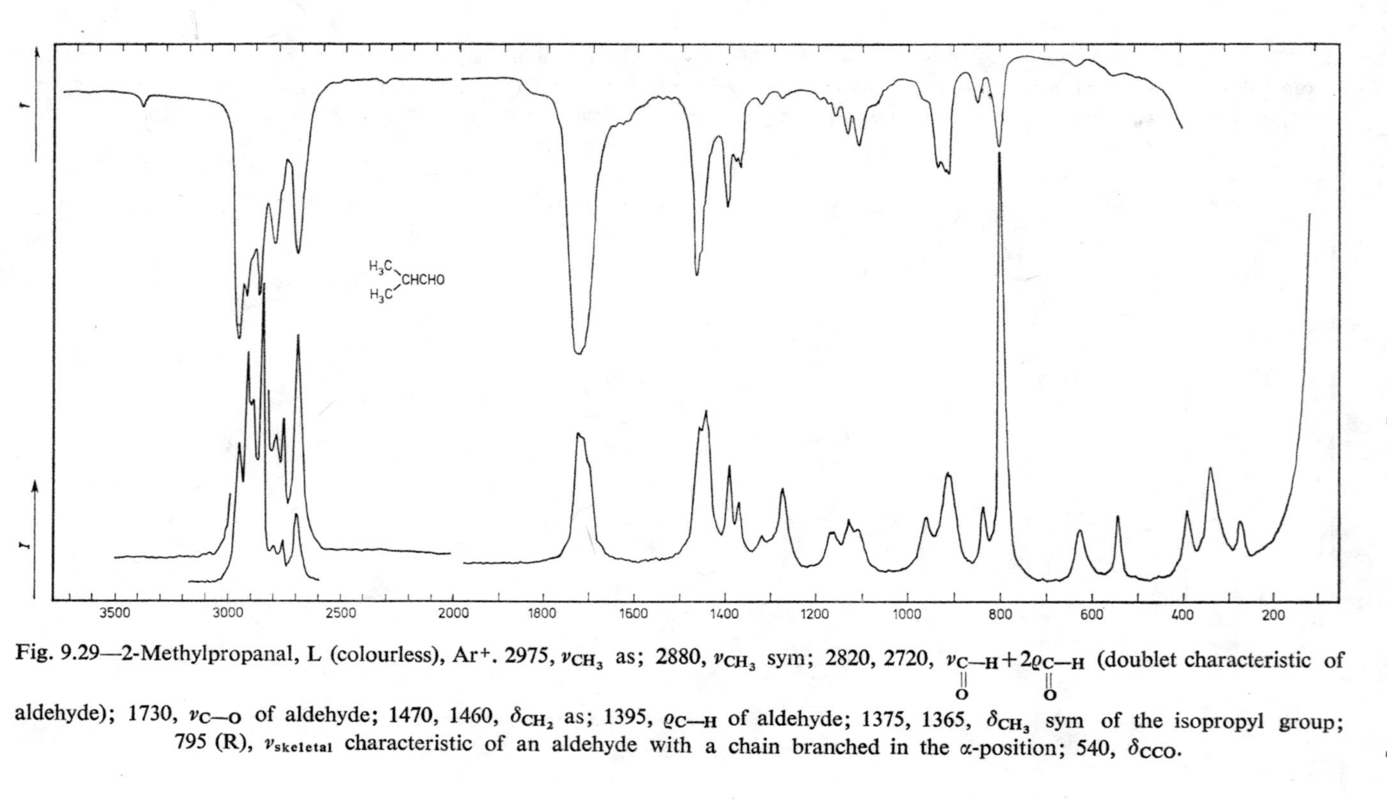

Fig. 9.29—2-Methylpropanal, L (colourless), Ar+. 2975, ν_{CH_3} as; 2880, ν_{CH_3} sym; 2820, 2720, $\nu_{\underset{\|}{C}-H}$ + $2\varrho_{\underset{\|}{C}-H}$ (doublet characteristic of aldehyde); 1730, $\nu_{C=O}$ of aldehyde; 1470, 1460, δ_{CH_2} as; 1395, ϱ_{C-H} of aldehyde; 1375, 1365, δ_{CH_3} sym of the isopropyl group; 795 (R), $\nu_{skeletal}$ characteristic of an aldehyde with a chain branched in the α-position; 540, δ_{CCO}.

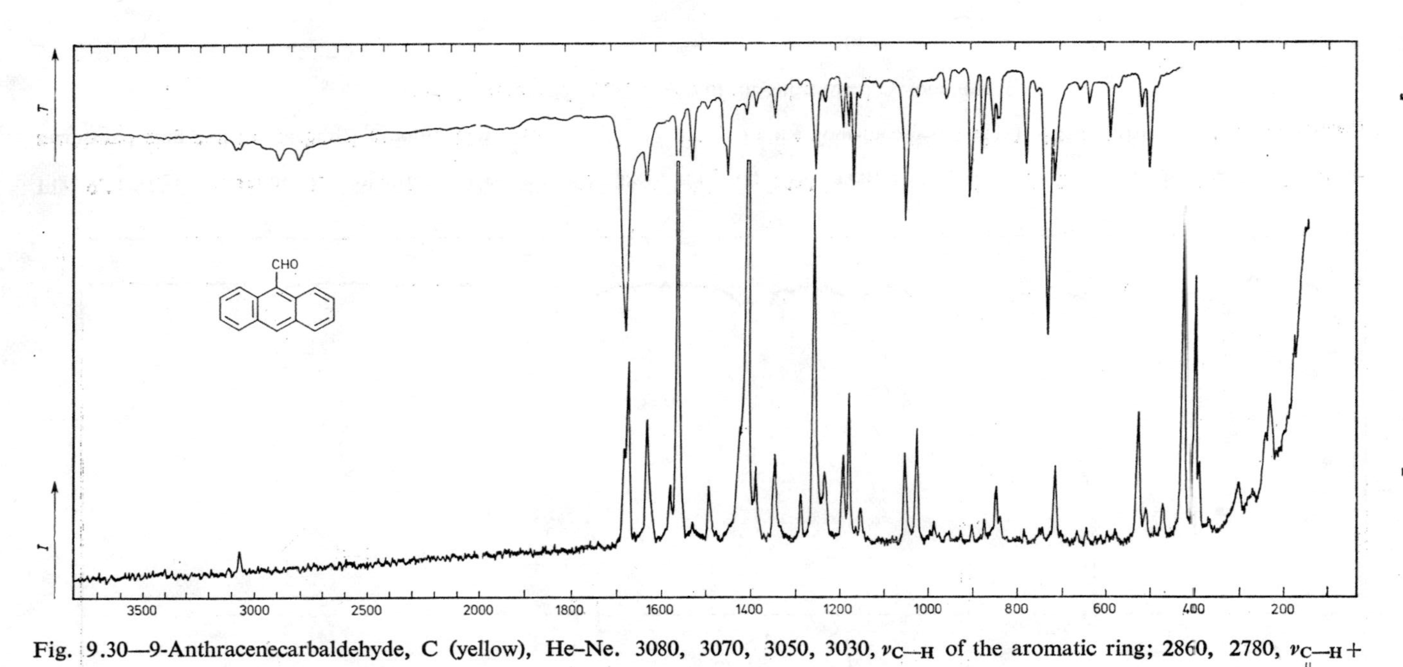

Fig. 9.30—9-Anthracenecarbaldehyde, C (yellow), He–Ne. 3080, 3070, 3050, 3030, $\nu_{\mathrm{C-H}}$ of the aromatic ring; 2860, 2780, $\nu_{\underset{\underset{\mathrm{O}}{\|}}{\mathrm{C}}\mathrm{-H}}+$ $+2\varrho_{\underset{\underset{\mathrm{O}}{\|}}{\mathrm{C}}\mathrm{-H}}$ (doublet characteristic of aldehyde); 1675, $\nu_{\mathrm{C=O}}$ (the reason for this particularly low frequency is not clear; 1625, 1555, 1400, $\nu_{\mathrm{C=C}}$ of the anthracene ring; 1390, $\varrho_{\underset{\underset{\mathrm{O}}{\|}}{\mathrm{C}}\mathrm{-H}}$; 1250, $\nu_{\mathrm{C-C}}$ of the aromatic aldehyde; 1055, 1020, $\delta_{\mathrm{C-H}}$; 735 (IR), $\gamma_{\mathrm{C-H}}$ of the four neighbouring hydrogen atoms; 425, 420, $\delta_{\mathrm{C-C}}$ of anthracene.

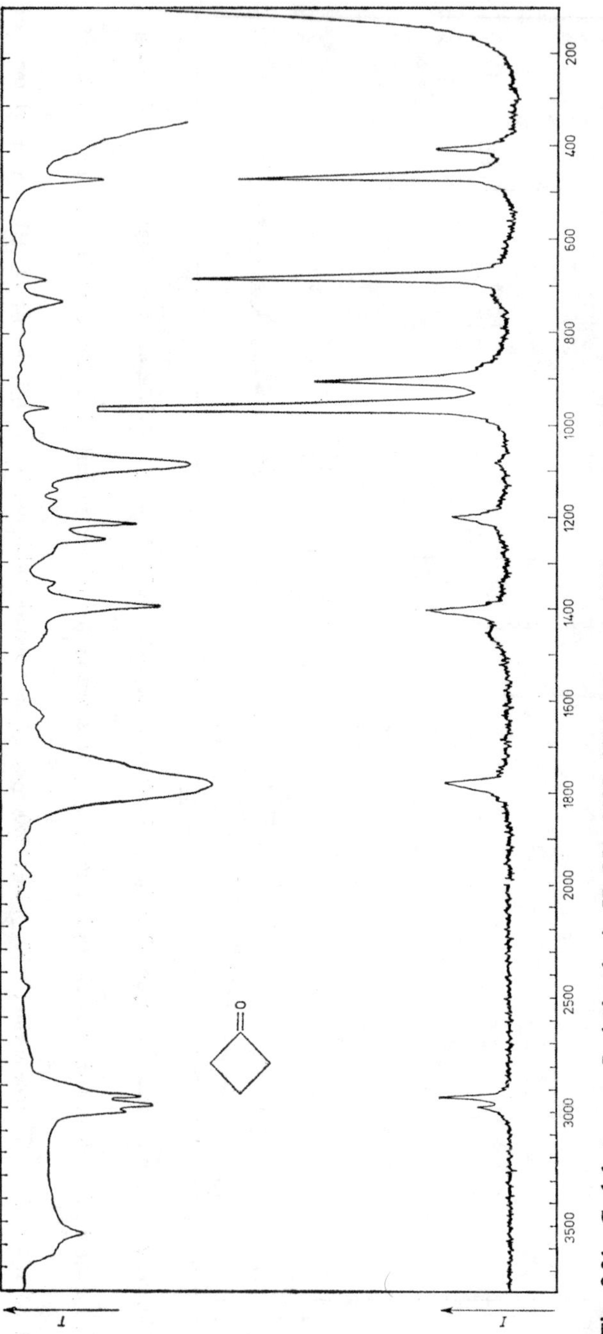

Fig. 9.31—Cyclobutanone, **L** (colourless), He–Ne. 3005, 2980, ν_{CH_2} as; 2930, ν_{CH_2} sym (this high frequency is due to the four-membered ring and the carbonyl group); 1780, $\nu_{C=O}$ ($>$C=O group in the four-membered ring); 1240, 1210, ω_{CH_2}; 1080, $\nu_{skeletal}$ as of the ring; 955, $\nu_{skeletal}$ sym of the substituted cyclobutane.

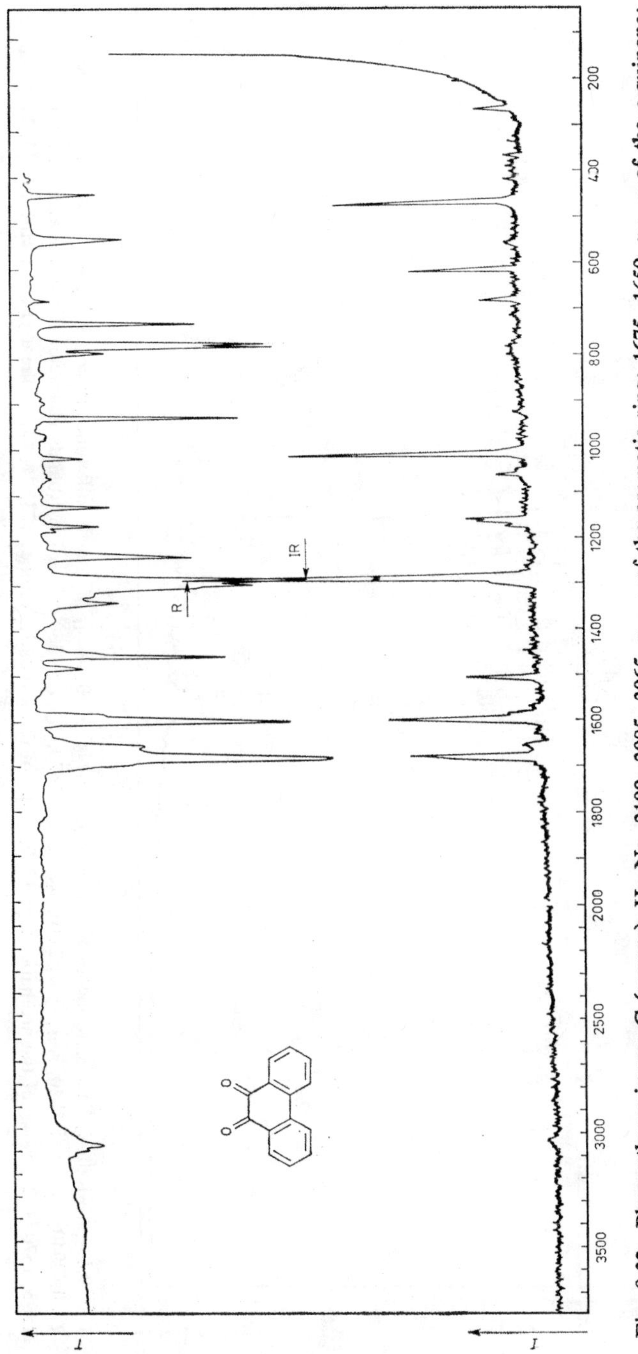

Fig. 9.32—Phenanthraquinone, C (orange), He–Ne. 3100, 3085, 3065, ν_{C-H} of the aromatic ring; 1675, 1650, $\nu_{C=O}$ of the *o*-quinone; 1590, 1500, 1450, 1280, $\nu_{C=C}$ of the phenanthrene structure; 770, 765, γ_{C-H} of the four neighbouring hydrogen atoms.

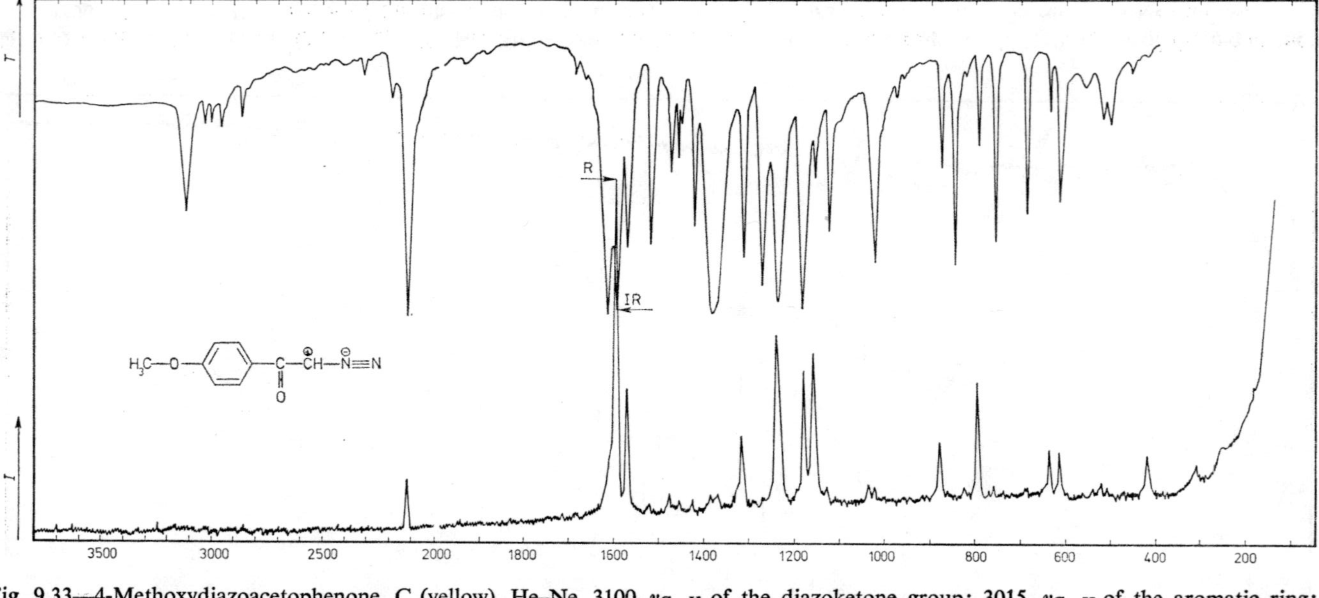

Fig. 9.33—4-Methoxydiazoacetophenone, C (yellow), He–Ne. 3100, ν_{C-H} of the diazoketone group; 3015, ν_{C-H} of the aromatic ring; 2980, 2940, ν_{CH_3} as; 2840, ν_{CH_3} sym; 2110, $\nu_{N=N}$ as; 1620, $\nu_{C=O}$; 1380, $\nu_{N=N}$ sym (bands characteristic of the aromatic diazoketone); 1595, 1570, 1520, $\nu_{C\equiv C}$ of the aromatic ring; 1310, ν_{C-C} of the aromatic ketone; 1235, ν_{C-O} of the aryl alkyl ether; 1175, 1155, δ_{C-H}; 840, γ_{C-H}; 790, $\nu_{C=C}$ of the *p*-substituted benzene.

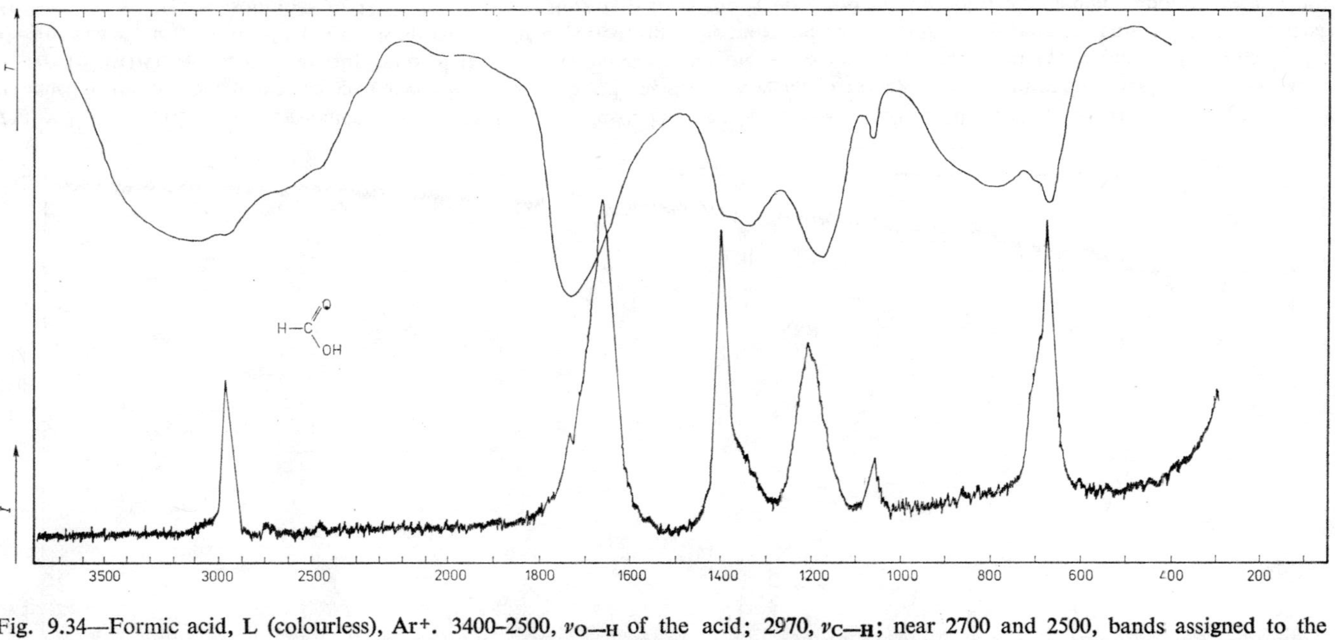

Fig. 9.34—Formic acid, L (colourless), Ar$^+$. 3400–2500, ν_{O-H} of the acid; 2970, ν_{C-H}; near 2700 and 2500, bands assigned to the combination tones characteristic of the acid; 1725 (IR), $\nu_{C=O}$ as of the acid dimer; 1660 (R), $\nu_{C=O}$ sym of the acid dimer; 1400, ϱ_{C-H}; 1335, δ_{O-H}; 1200 (R), ν_{C-O}; 1170, ν_{C-O}; 750, γ_{O-H}; 675, $\delta_{C=O}$.

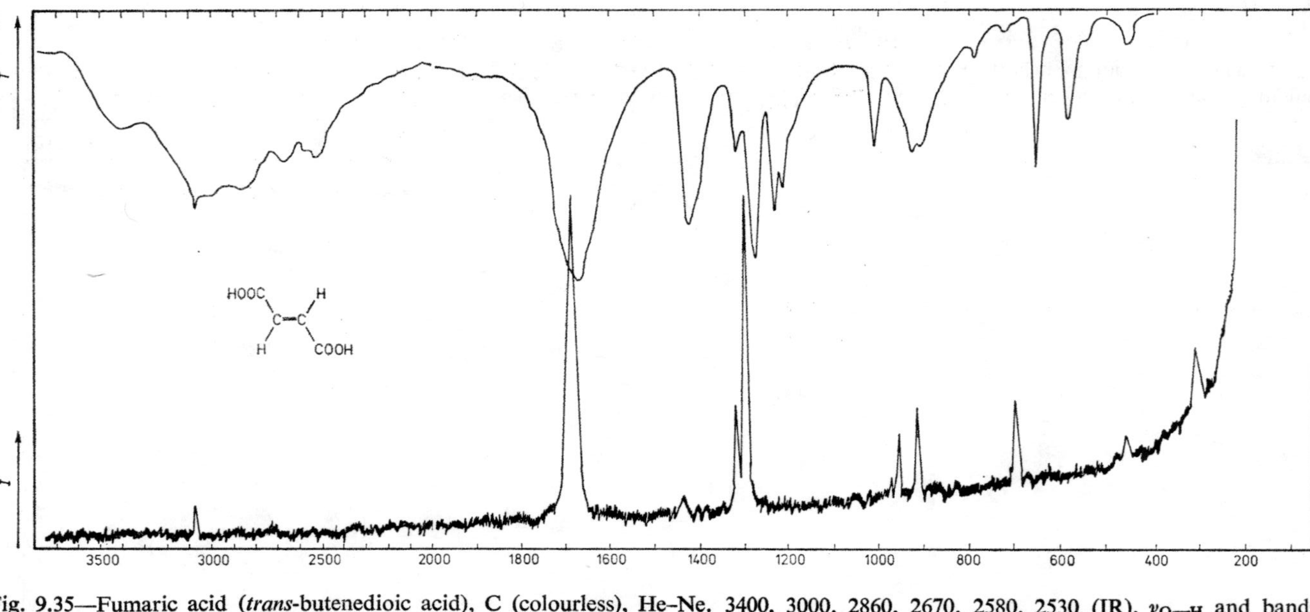

Fig. 9.35—Fumaric acid (*trans*-butenedioic acid), C (colourless), He–Ne. 3400, 3000, 2860, 2670, 2580, 2530 (IR), ν_{O-H} and bands assigned to the combination tones characteristic of the carboxylic acid; 3080, ν_{C-H} of the hydrogen atom at the double bond; 1685 (R), $\nu_{C=O}$, 1670 (IR), $\nu_{C=O}$, the intensity ratio of these bands indicates that the compound does not form cyclic dimers (cf. Fig. 9.34); the absence of the $\nu_{C=C}$ band from the R spectrum is unexpected, since the molecule has a centre of symmetry. This band appears near 1650 in the spectra of solutions in H_2O, CH_3OH or dioxan whereas the $\nu_{C=O}$ band of the solvated monomer appears near 1730; 1425, δ_{O-H}; 1275, ν_{C-O}; 925, γ_{C-H} of the *trans*-disubstituted ethylene; 900, γ_{O-H}; 645, δ_{OCO} of the acid.

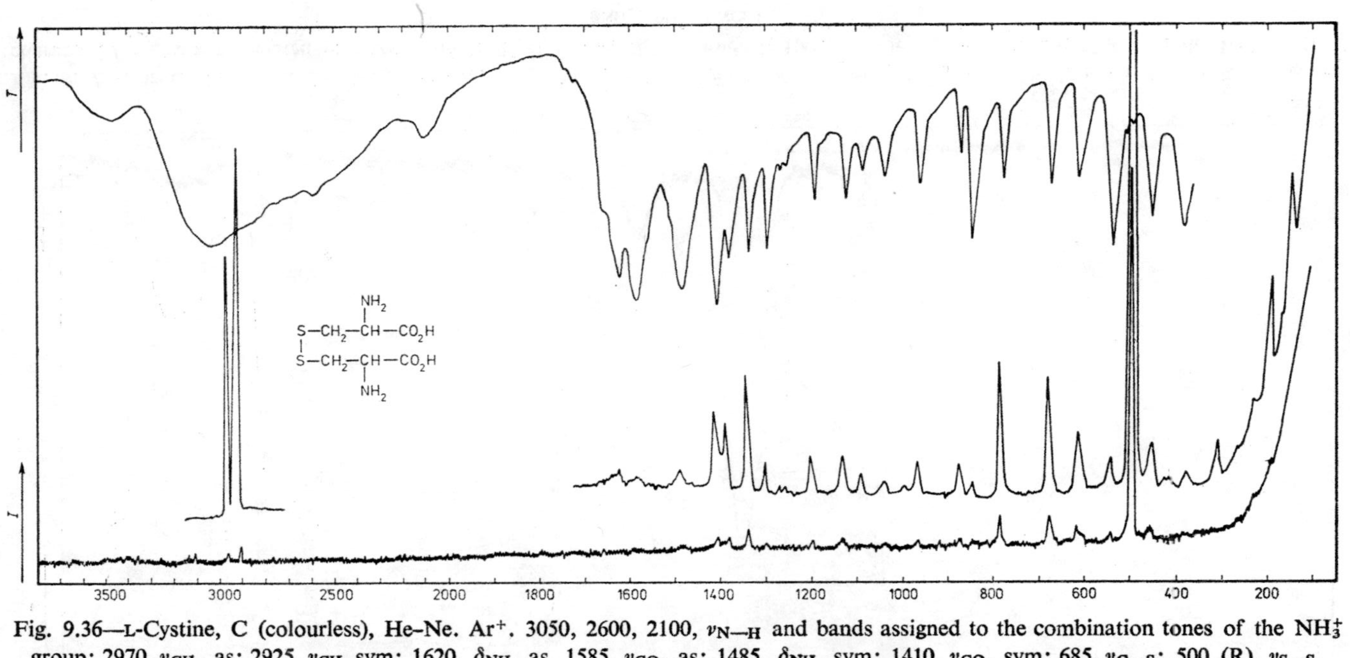

Fig. 9.36—L-Cystine, C (colourless), He–Ne. Ar^+. 3050, 2600, 2100, ν_{N-H} and bands assigned to the combination tones of the NH_3^+ group; 2970, ν_{CH_2} as; 2925, ν_{CH_2} sym; 1620, δ_{NH_3} as, 1585, ν_{CO_2} as; 1485, δ_{NH_3} sym; 1410, ν_{CO_2} sym; 685, ν_{C-S}; 500 (R), ν_{S-S}.

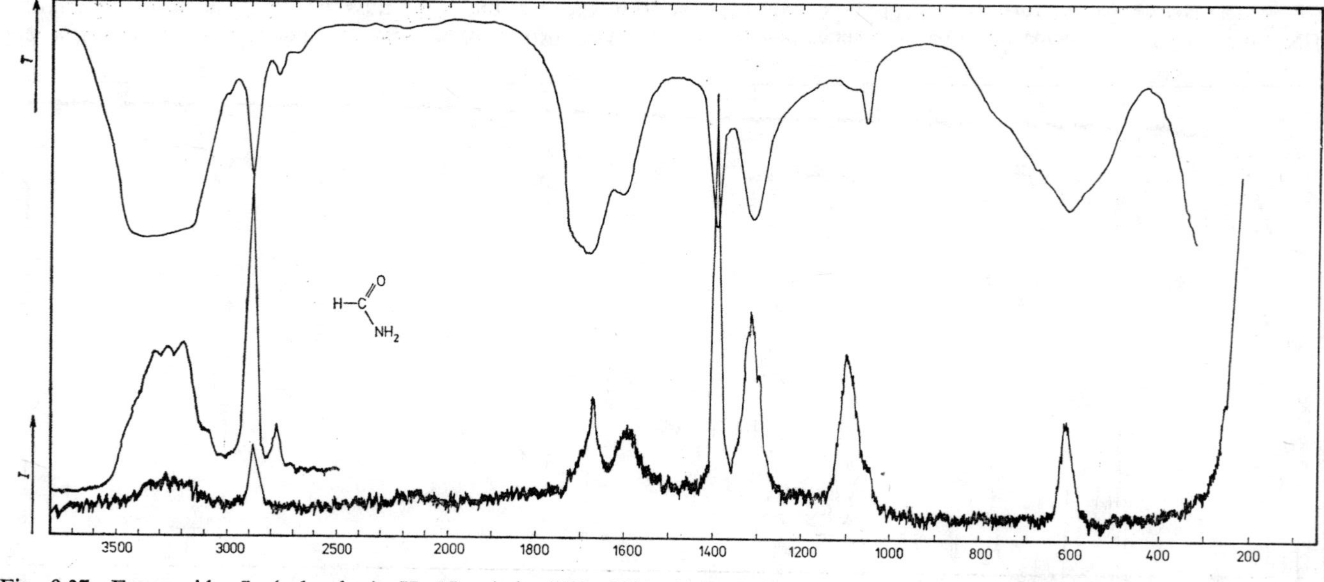

Fig. 9.37—Formamide, L (colourless), He–Ne. Ar^+. 3420, 3320, 3270, 3190 ν_{NH_2} as and ν_{NH_2} sym (the four bands in this region are characteristic of the associated primary amide); 2890, $\nu_{C—H}$; 1675, Amide I; 1600, Amide II; 1390, $\varrho_{C—H}$; 1310, Amide III; 1095 (R), ϱ_{NH_2}; 605, δ_{NCO}.

Fig. 9.38—*N*-Cyclohexylacetamide, C (colourless), He–Ne, Ar^+. 3290, ν_{N-H} 3085, overtone of the Amide II band; 1635, Amide I of the secondary amide; 1555, Amide II of the secondary amide; 1445, δ_{CH_2} of the ring; 1255, Amide III, 800, $\nu_{skeletal}$ of the six-membered ring; 795 (IR), γ_{NH_2}; 640, δ_{NCO};605 (IR), γ_{NCO}.

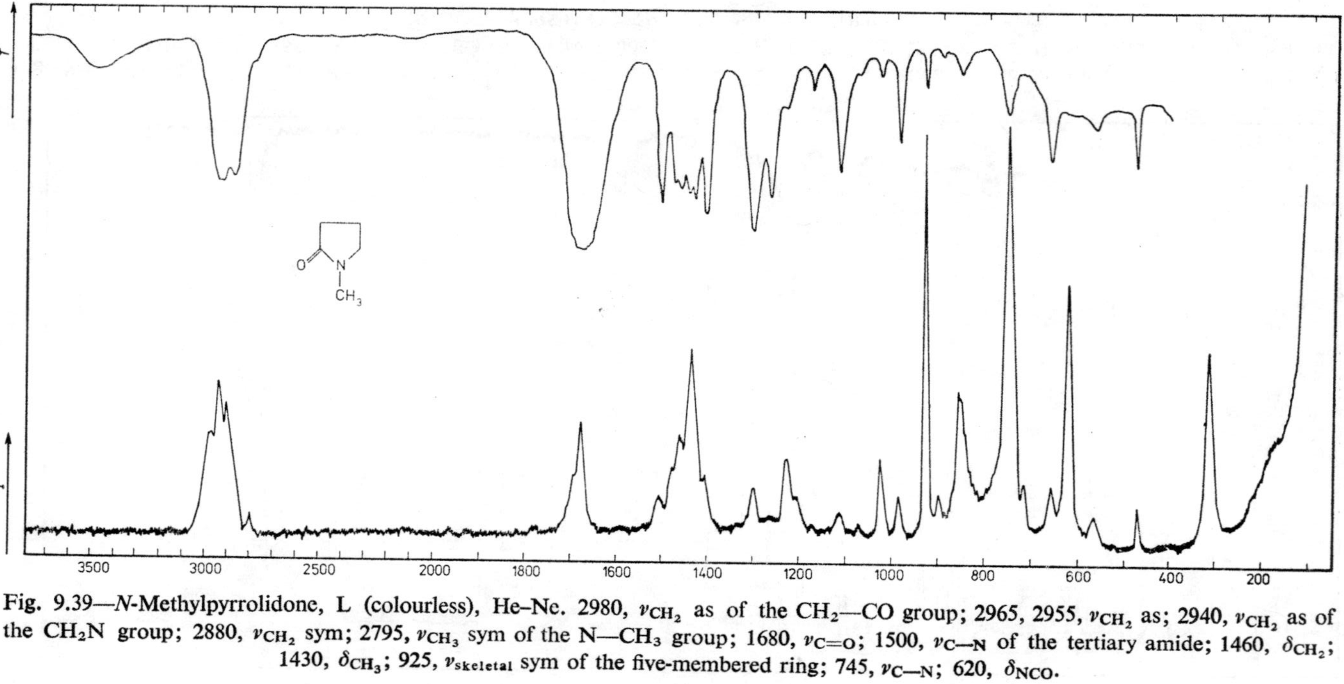

Fig. 9.39—*N*-Methylpyrrolidone, L (colourless), He–Ne. 2980, ν_{CH_2} as of the CH_2—CO group; 2965, 2955, ν_{CH_2} as; 2940, ν_{CH_2} as of the CH_2N group; 2880, ν_{CH_2} sym; 2795, ν_{CH_3} sym of the N—CH_3 group; 1680, $\nu_{C=O}$; 1500, $\nu_{C—N}$ of the tertiary amide; 1460, δ_{CH_2}; 1430, δ_{CH_3}; 925, $\nu_{skeletal}$ sym of the five-membered ring; 745, $\nu_{C—N}$; 620, δ_{NCO}.

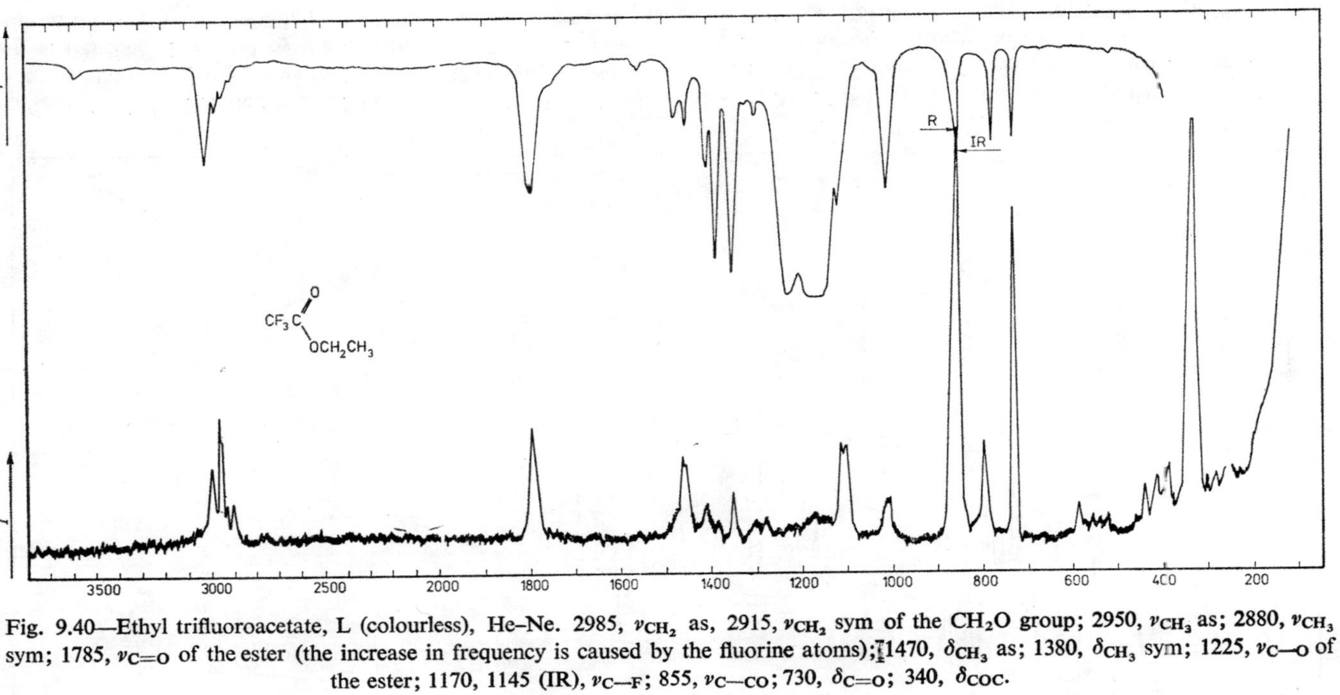

Fig. 9.40—Ethyl trifluoroacetate, L (colourless), He–Ne. 2985, ν_{CH_2} as, 2915, ν_{CH_2} sym of the CH_2O group; 2950, ν_{CH_3} as; 2880, ν_{CH_3} sym; 1785, $\nu_{C=O}$ of the ester (the increase in frequency is caused by the fluorine atoms); 1470, δ_{CH_3} as; 1380, δ_{CH_3} sym; 1225, ν_{C-O} of the ester; 1170, 1145 (IR), ν_{C-F}; 855, ν_{C-CO}; 730, $\delta_{C=O}$; 340, δ_{COC}.

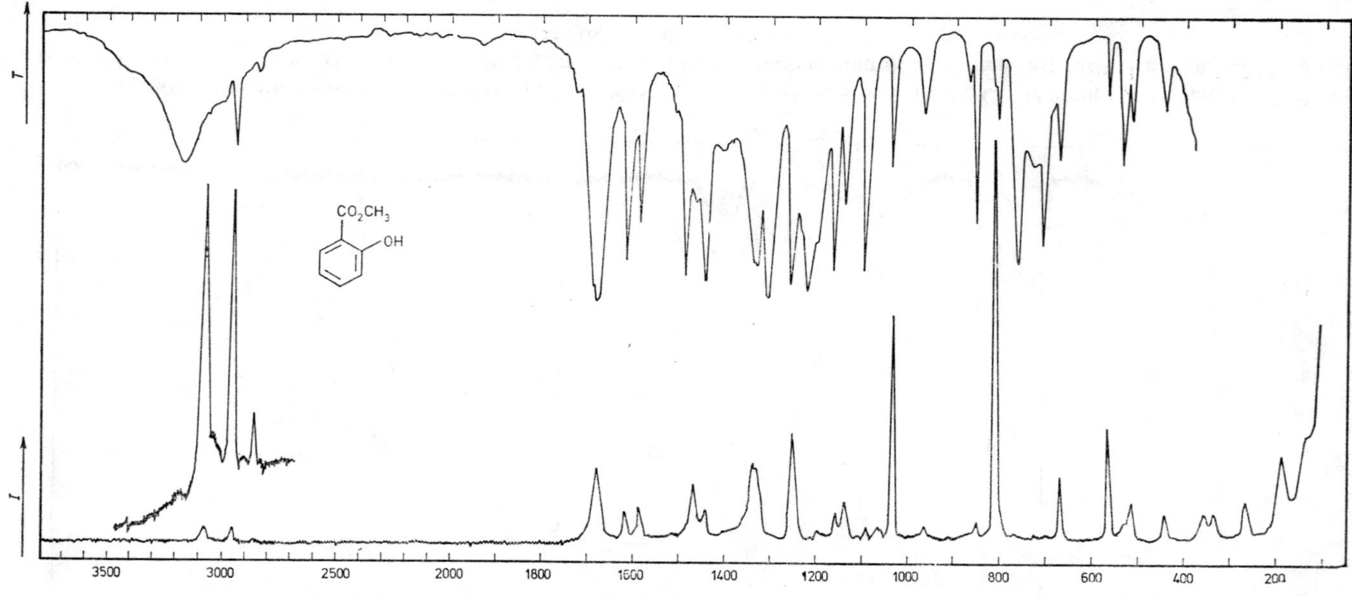

Fig. 9.41—Methyl salicylate (methyl 2-hydroxybenzoate), L (colourless), He–Ne. Ar^+. 3180, ν_{O-H} of the intramolecularly bonded OH group; 3080, ν_{C-H} of the aromatic ring; 2965, 2860, ν_{CH_3} as and ν_{CH_3} sym of the OCH_3 group; 1675, $\nu_{C=O}$ of the aromatic ester (the lowered frequency is caused by the participation of the C=O group in the intramolecular hydrogen bond); 1615, 1585, 1485, $\nu_{C=C}$ of the aromatic ring; 1460, 1440, δ_{CH_3}; 1305, 1155, ν_{C-O} of the aromatic ester; 1205, ν_{C-O} of the phenolic group; 1035, δ_{C-H}, 755, γ_{C-H} of the *o*-substituted benzene.

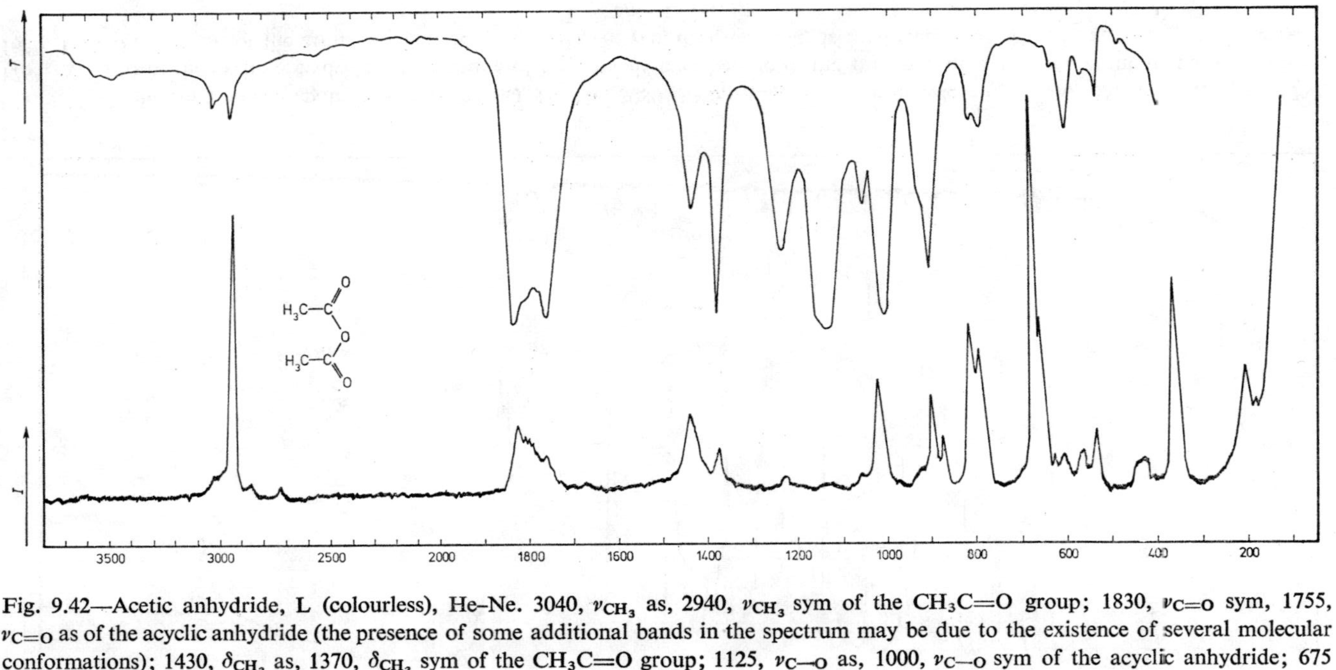

Fig. 9.42—Acetic anhydride, L (colourless), He–Ne. 3040, ν_{CH_3} as, 2940, ν_{CH_3} sym of the $CH_3C{=}O$ group; 1830, $\nu_{C=O}$ sym, 1755, $\nu_{C=O}$ as of the acyclic anhydride (the presence of some additional bands in the spectrum may be due to the existence of several molecular conformations); 1430, δ_{CH_3} as, 1370, δ_{CH_3} sym of the $CH_3C{=}O$ group; 1125, ν_{C-O} as, 1000, ν_{C-O} sym of the acyclic anhydride; 675 (R), $\delta_{C=O}$.

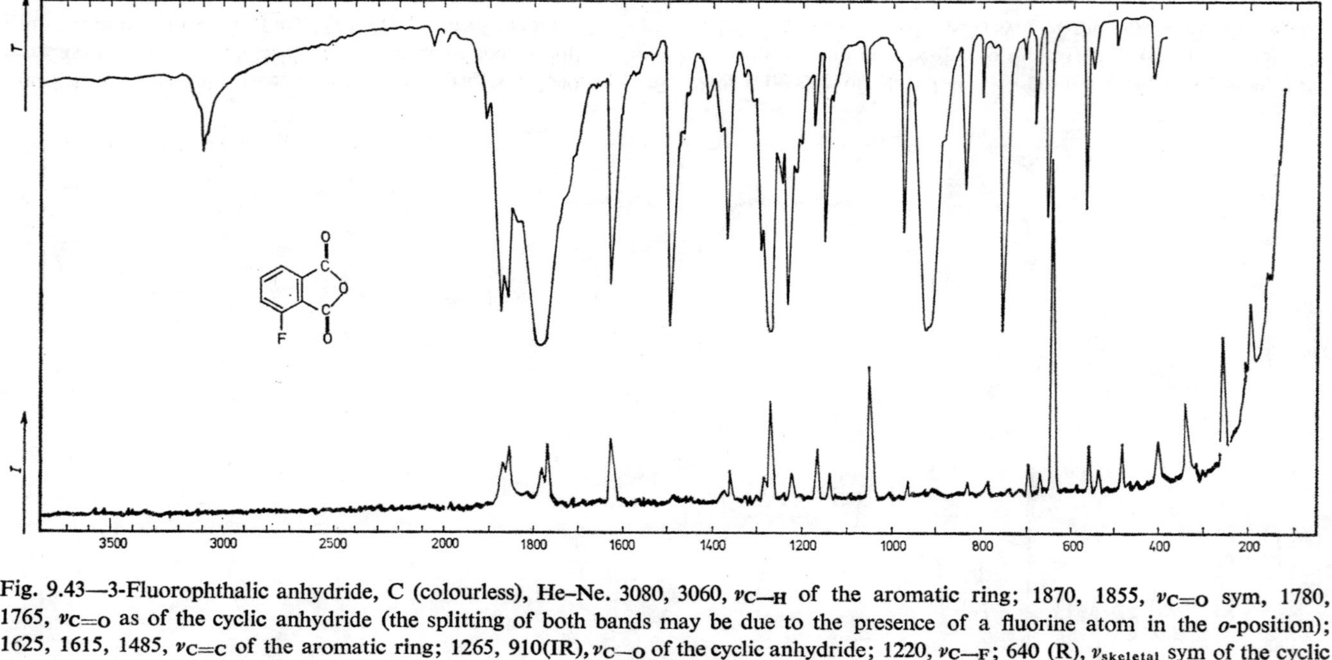

Fig. 9.43—3-Fluorophthalic anhydride, C (colourless), He–Ne. 3080, 3060, ν_{C-H} of the aromatic ring; 1870, 1855, $\nu_{C=O}$ sym, 1780, 1765, $\nu_{C=O}$ as of the cyclic anhydride (the splitting of both bands may be due to the presence of a fluorine atom in the *o*-position); 1625, 1615, 1485, $\nu_{C=C}$ of the aromatic ring; 1265, 910(IR), ν_{C-O} of the cyclic anhydride; 1220, ν_{C-F}; 640 (R), $\nu_{skeletal}$ sym of the cyclic anhydride.

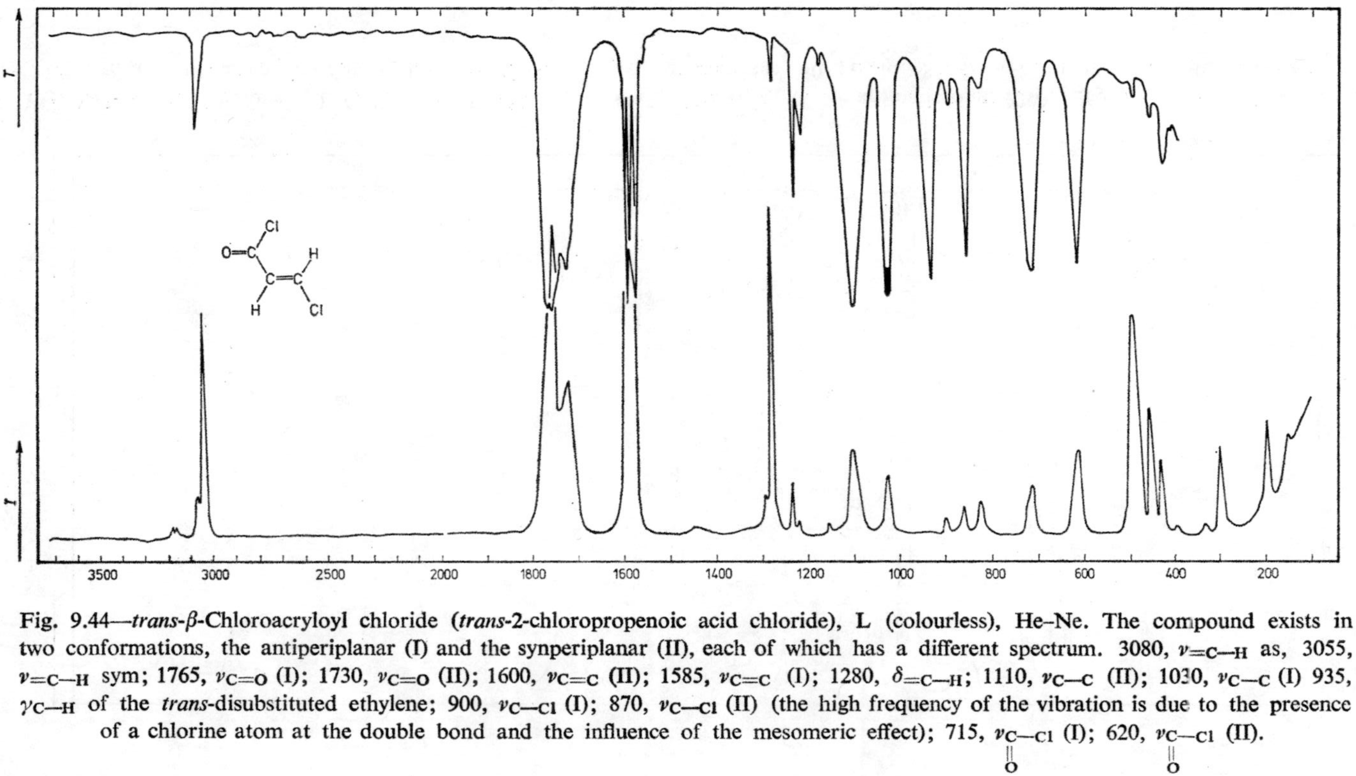

Fig. 9.44—*trans*-β-Chloroacryloyl chloride (*trans*-2-chloropropenoic acid chloride), **L** (colourless), He–Ne. The compound exists in two conformations, the antiperiplanar (I) and the synperiplanar (II), each of which has a different spectrum. 3080, $\nu_{=C-H}$ as, 3055, $\nu_{=C-H}$ sym; 1765, $\nu_{C=O}$ (I); 1730, $\nu_{C=O}$ (II); 1600, $\nu_{C=C}$ (II); 1585, $\nu_{C=C}$ (I); 1280, $\delta_{=C-H}$; 1110, ν_{C-C} (II); 1030, ν_{C-C} (I) 935, γ_{C-H} of the *trans*-disubstituted ethylene; 900, ν_{C-Cl} (I); 870, ν_{C-Cl} (II) (the high frequency of the vibration is due to the presence of a chlorine atom at the double bond and the influence of the mesomeric effect); 715, $\nu_{\underset{\parallel}{\overset{}{C}}-Cl}$ with C=O (I); 620, ν_{C-Cl} with C=O (II).

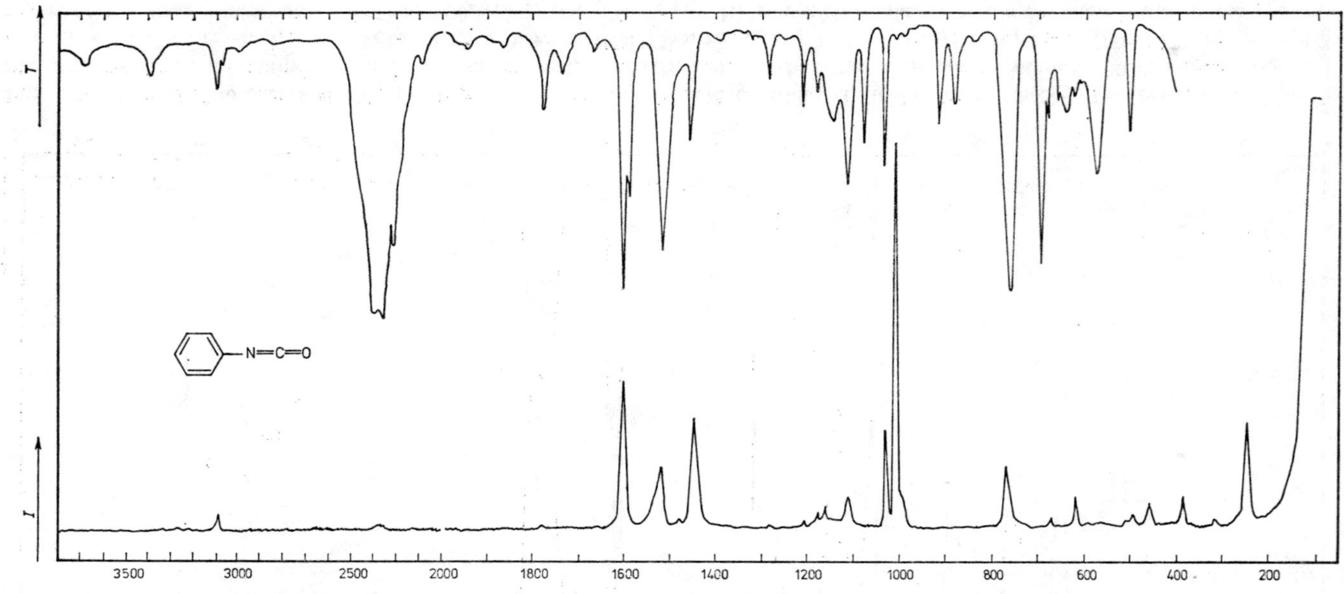

Fig. 9.45—Phenyl isocyanate, L (colourless), He–Ne. 3075, 3065, 3030, $\nu_{C—H}$ of the aromatic ring; 2290, 2260, $\nu_{N=C=O}$ as; 1600, 1585, 1510, $\nu_{C=C}$ of the aromatic ring; 1030, $\delta_{C—H}$, 1005, $\nu_{C=C}$ sym of the ring; 750, $\gamma_{C—H}$, 685, $\delta_{C—C}$ of the monosubstituted benzene.

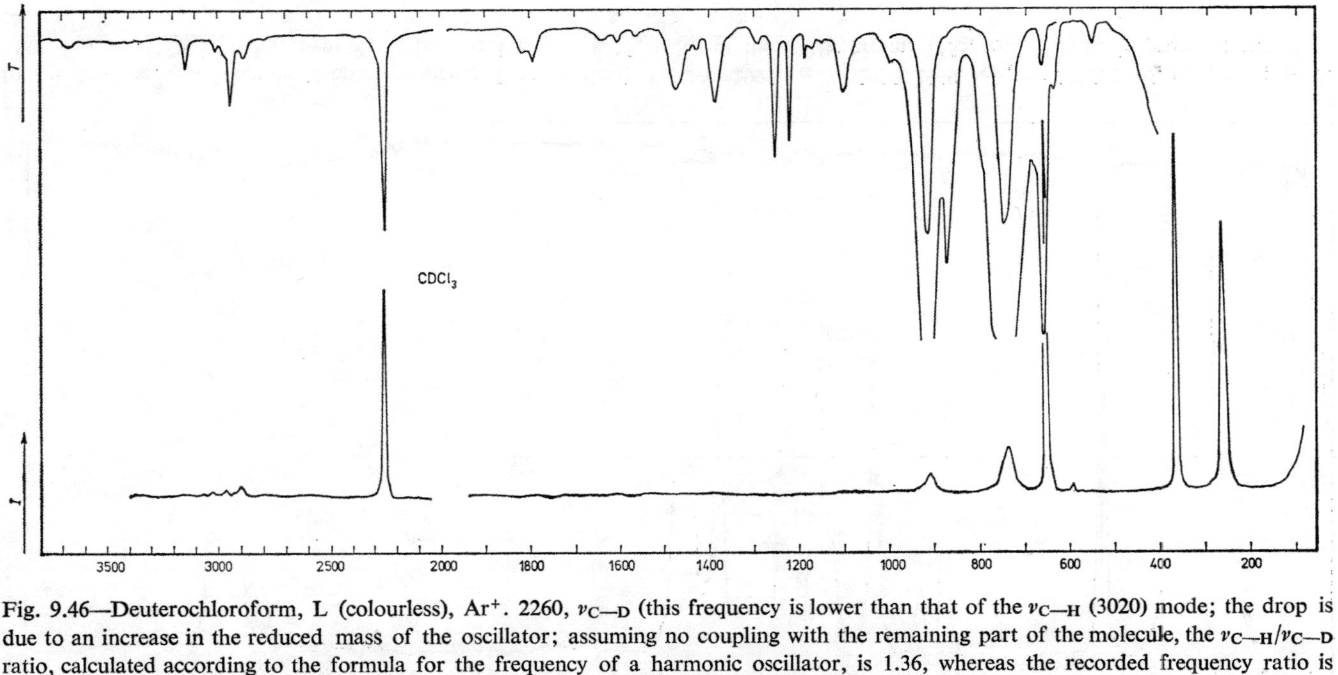

Fig. 9.46—Deuterochloroform, L (colourless), Ar$^+$. 2260, ν_{C-D} (this frequency is lower than that of the ν_{C-H} (3020) mode; the drop is due to an increase in the reduced mass of the oscillator; assuming no coupling with the remaining part of the molecule, the ν_{C-H}/ν_{C-D} ratio, calculated according to the formula for the frequency of a harmonic oscillator, is 1.36, whereas the recorded frequency ratio is 1.34); 910, δ_{C-D}; 740, ν_{C-Cl} as; 650, ν_{C-Cl} sym; 335, 225, δ_{C-Cl}.

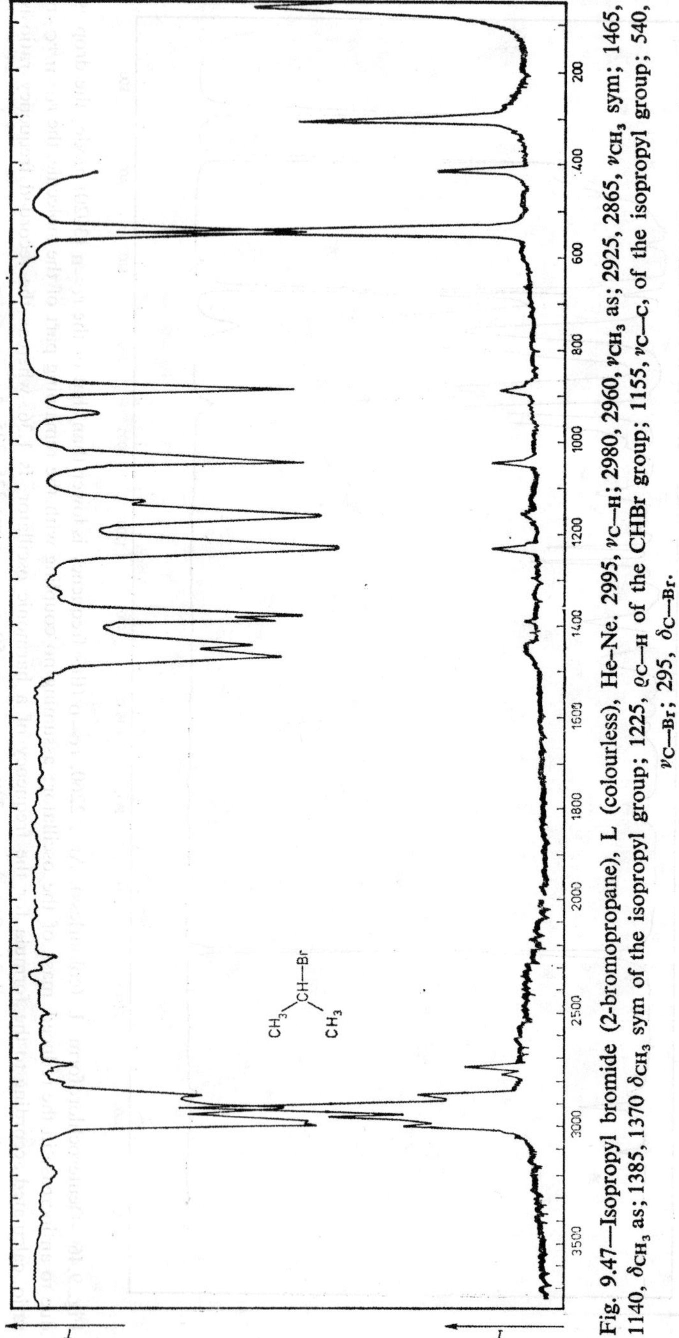

Fig. 9.47—Isopropyl bromide (2-bromopropane), L (colourless), He–Ne. 2995, $\nu_{C—H}$; 2980, 2960, ν_{CH_3} as; 2925, 2865, ν_{CH_3} sym; 1465, 1140, δ_{CH_3} as; 1385, 1370 δ_{CH_3} sym of the isopropyl group; 1225, $\varrho_{C—H}$ of the CHBr group; 1155, $\nu_{C—C}$, of the isopropyl group; 540, $\nu_{C—Br}$; 295, $\delta_{C—Br}$.

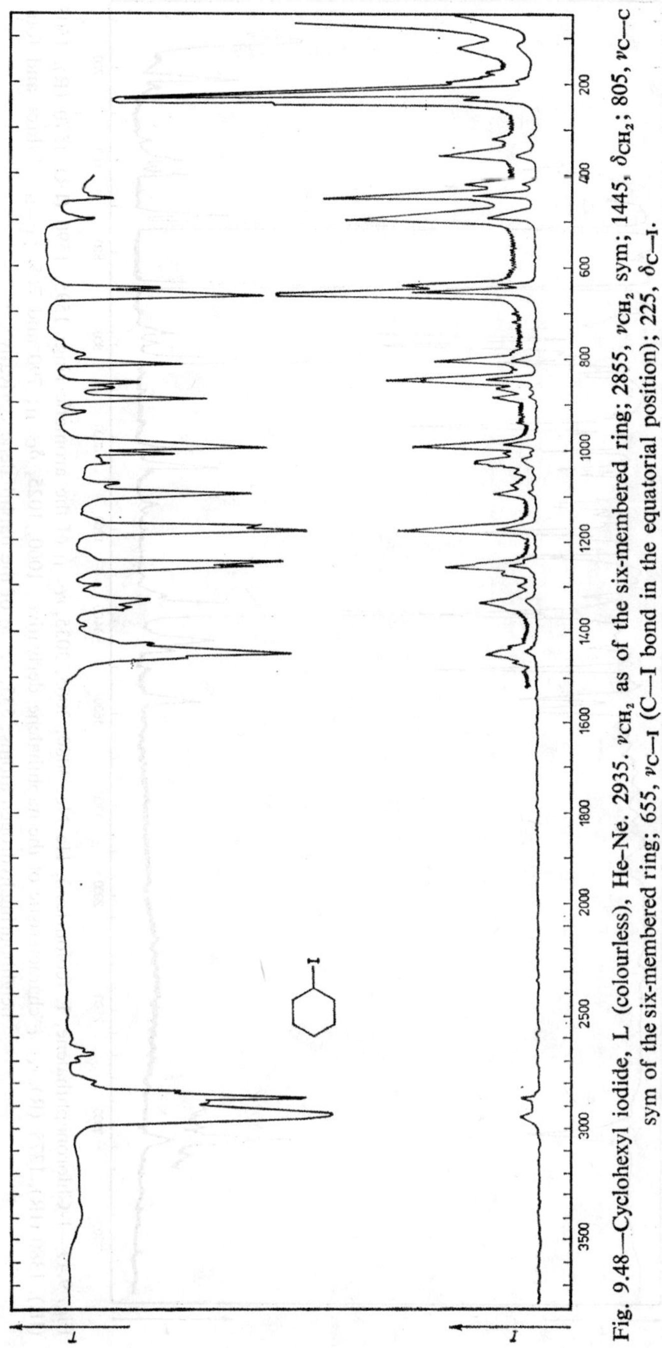

Fig. 9.48—Cyclohexyl iodide, L (colourless), He–Ne. 2935. ν_{CH_2} as of the six-membered ring; 2855, ν_{CH_2} sym; 1445, δ_{CH_2}; 805, $\nu_{C—C}$ sym of the six-membered ring; 655, $\nu_{C—I}$ (C—I bond in the equatorial position); 225, $\delta_{C—I}$.

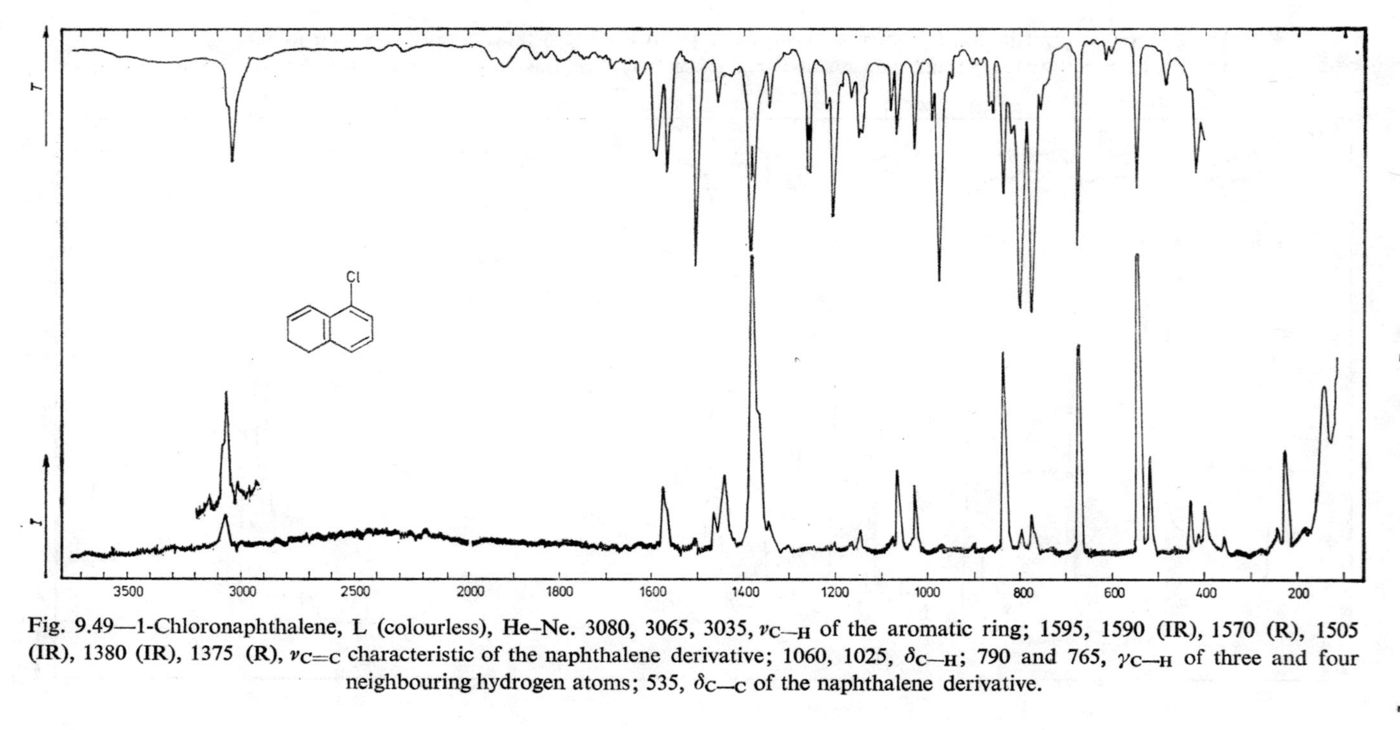

Fig. 9.49—1-Chloronaphthalene, L (colourless), He–Ne. 3080, 3065, 3035, ν_{C-H} of the aromatic ring; 1595, 1590 (IR), 1570 (R), 1505 (IR), 1380 (IR), 1375 (R), $\nu_{C=C}$ characteristic of the naphthalene derivative; 1060, 1025, δ_{C-H}; 790 and 765, γ_{C-H} of three and four neighbouring hydrogen atoms; 535, δ_{C-C} of the naphthalene derivative.

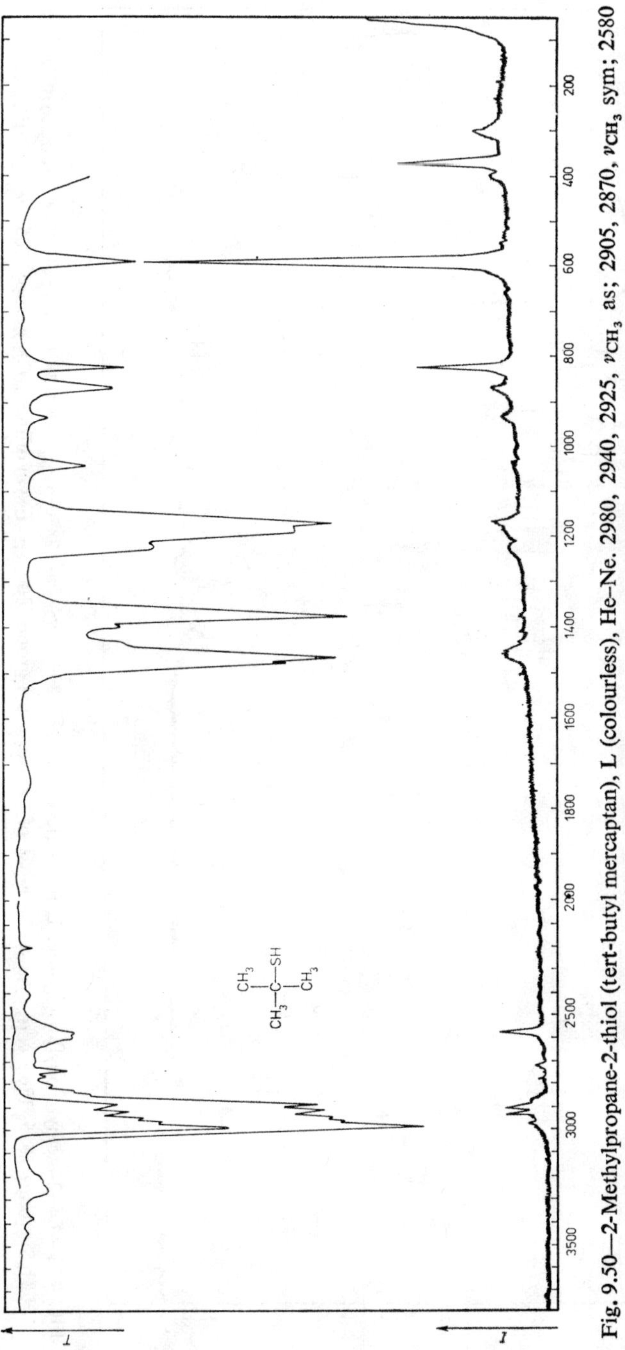

Fig. 9.50—2-Methylpropane-2-thiol (tert-butyl mercaptan), L (colourless), He–Ne. 2980, 2940, 2925, ν_{CH_3} as; 2905, 2870, ν_{CH_3} sym; 2580 (R), $\nu_{S—H}$; 1465, 1455, δ_{CH_3} as; 1390, 1375, δ_{CH_3} sym of the tert-butyl group; 1180, 1160, $\nu_{C—C}$ of the $C(C)_3$ group; 595, $\nu_{C—S}$ (IR spectrum of a capillary film and a layer of l = 0.02 mm).

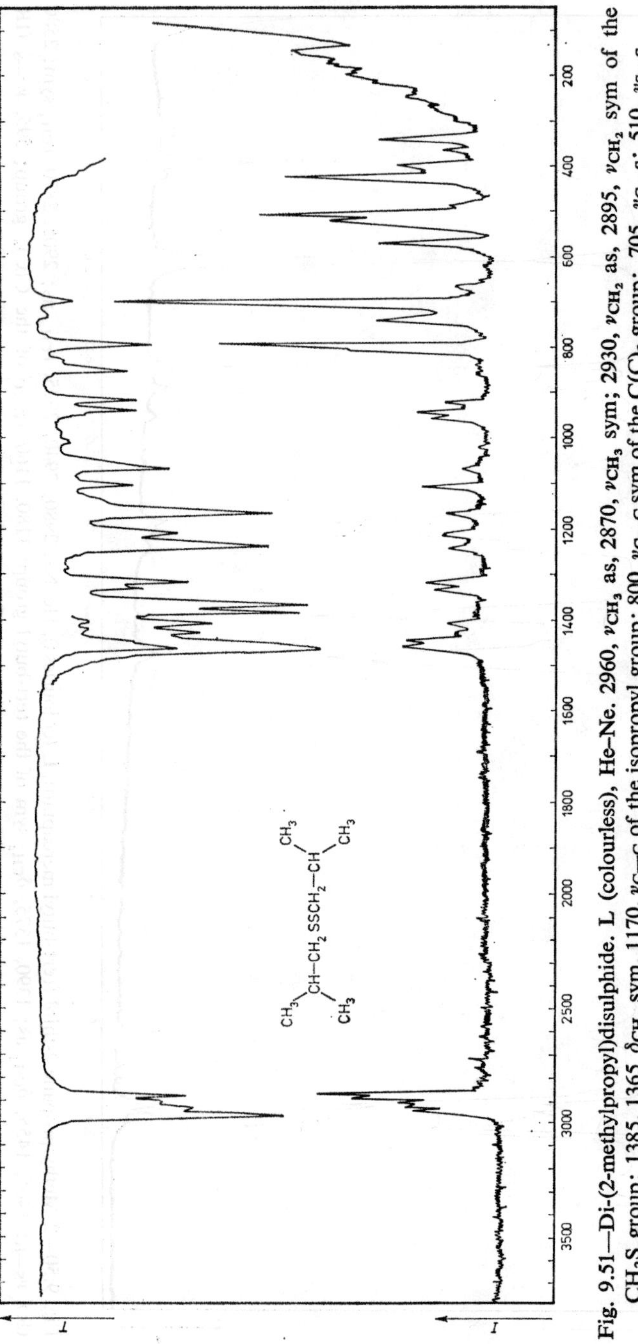

Fig. 9.51—Di-(2-methylpropyl)disulphide. L (colourless), He–Ne. 2960, ν_{CH_3} as, 2870, ν_{CH_3} sym; 2930, ν_{CH_2} as, 2895, ν_{CH_2} sym of the CH_2S group; 1385, 1365 δ_{CH_3} sym, 1170, $\nu_{C—C}$ of the isopropyl group; 800, $\nu_{C—C}$ sym of the $C(C)_3$ group;, 705, $\nu_{C—S}$; 510, $\nu_{S—S}$.

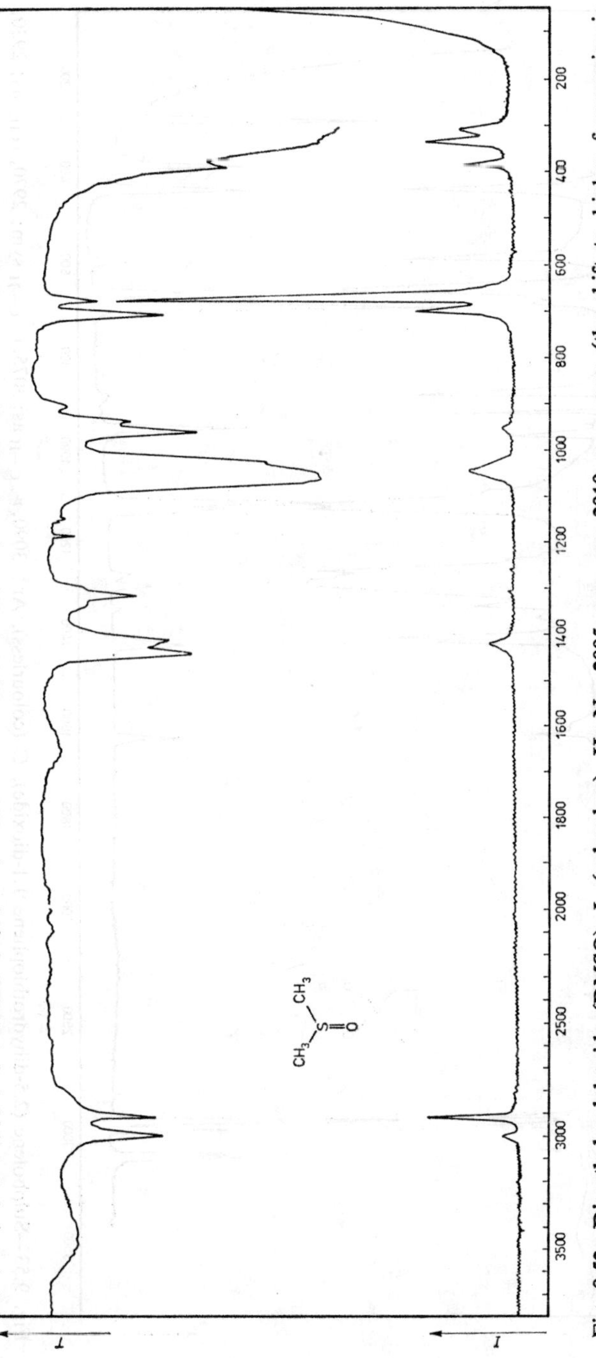

Fig. 9.52—Dimethyl sulphoxide (DMSO), L (colourless), He–Ne. 2995, ν_{CH_3} as; 2910, ν_{CH_3} sym (the shift to higher frequencies is caused by the sulphur atom); 1430, δ_{CH_3} as, 1300, δ_{CH_3} sym of the CH_3S group; 1034, $\nu_{S=O}$; 700, ν_{C-S} as; 670 (R), ν_{C-S} sym.

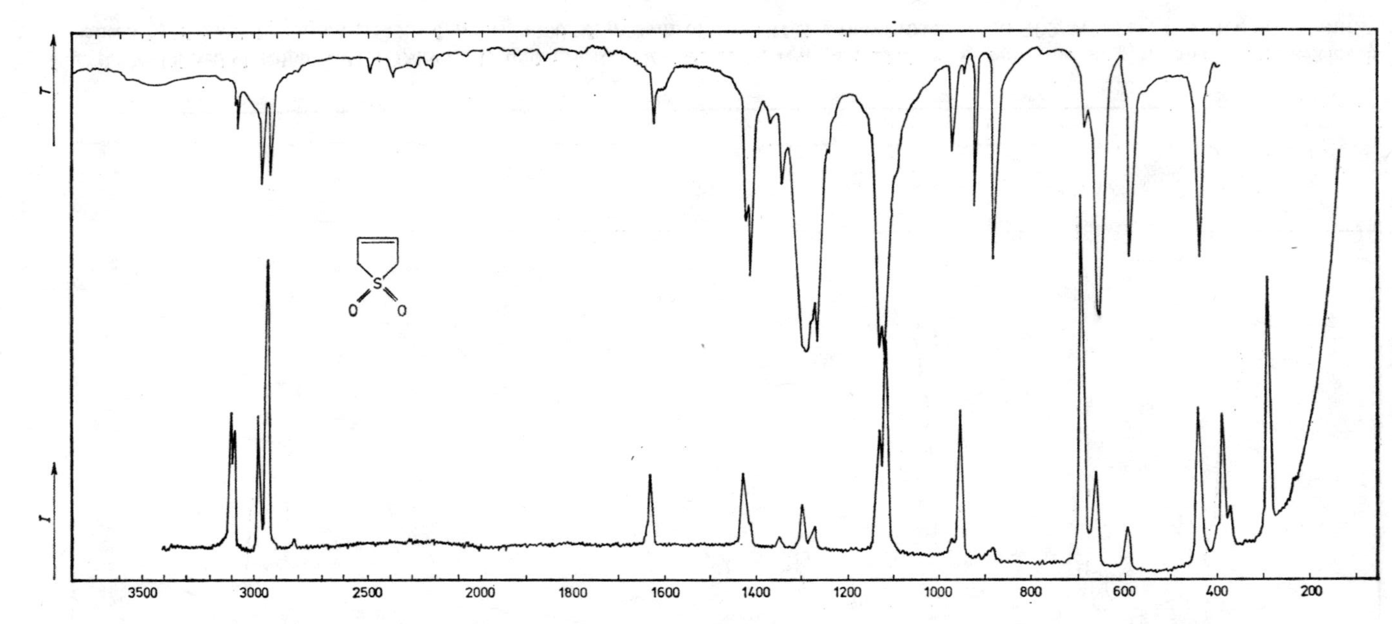

Fig. 9.53—Sulpholene (2,5-dihydrothiophene 1,1-dioxide), C (colourless), Ar^+ 3090, $\nu_{=C-H}$ as; 3075. $\nu_{=C-H}$ sym; 2970, ν_{CH_2} as; 2930, ν_{CH_2} sym; 1630, $\nu_{C=C}$; 1420, 1410, δ_{CH_2}; 1285, ν_{SO_2} as; 1130, 1120, ν_{SO_2} sym; 690, 660, ν_{CSC} sym and as; 585, δ_{SO_2}.

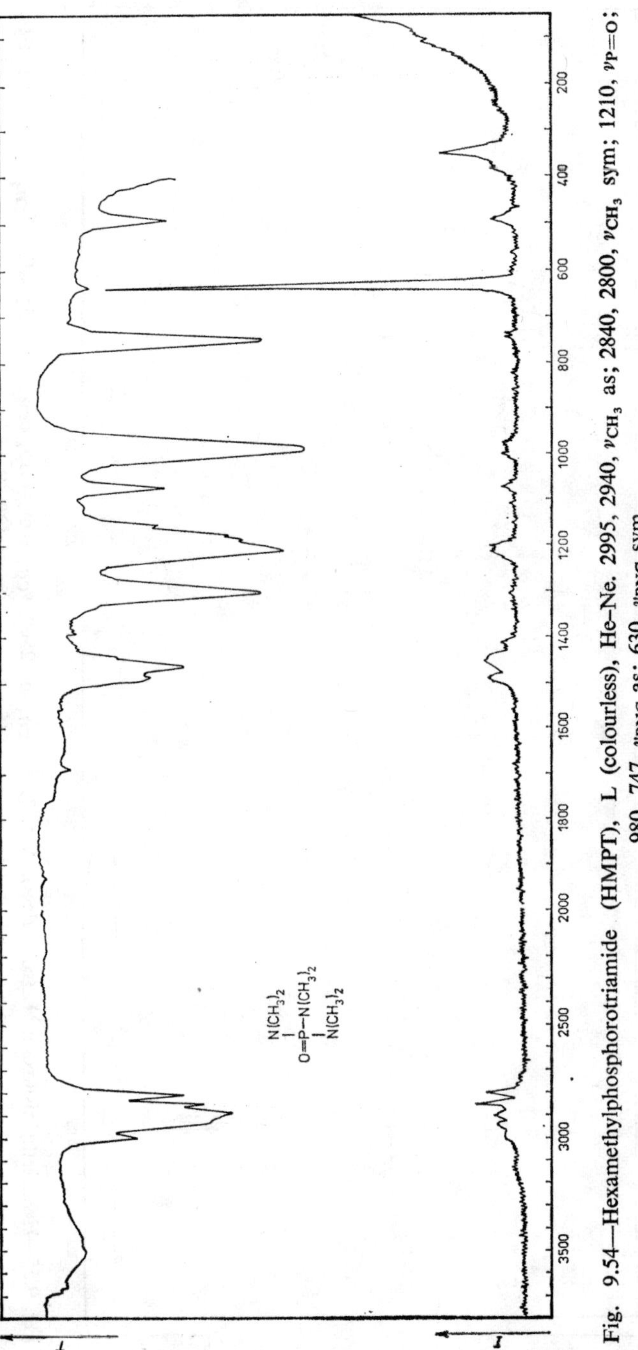

Fig. 9.54—Hexamethylphosphorotriamide (HMPT), L (colourless), He–Ne. 2995, 2940, ν_{CH_3} as; 2840, 2800, ν_{CH_3} sym; 1210, $\nu_{P=O}$; 980, 747, ν_{PNC} as; 630, ν_{PNC} sym.

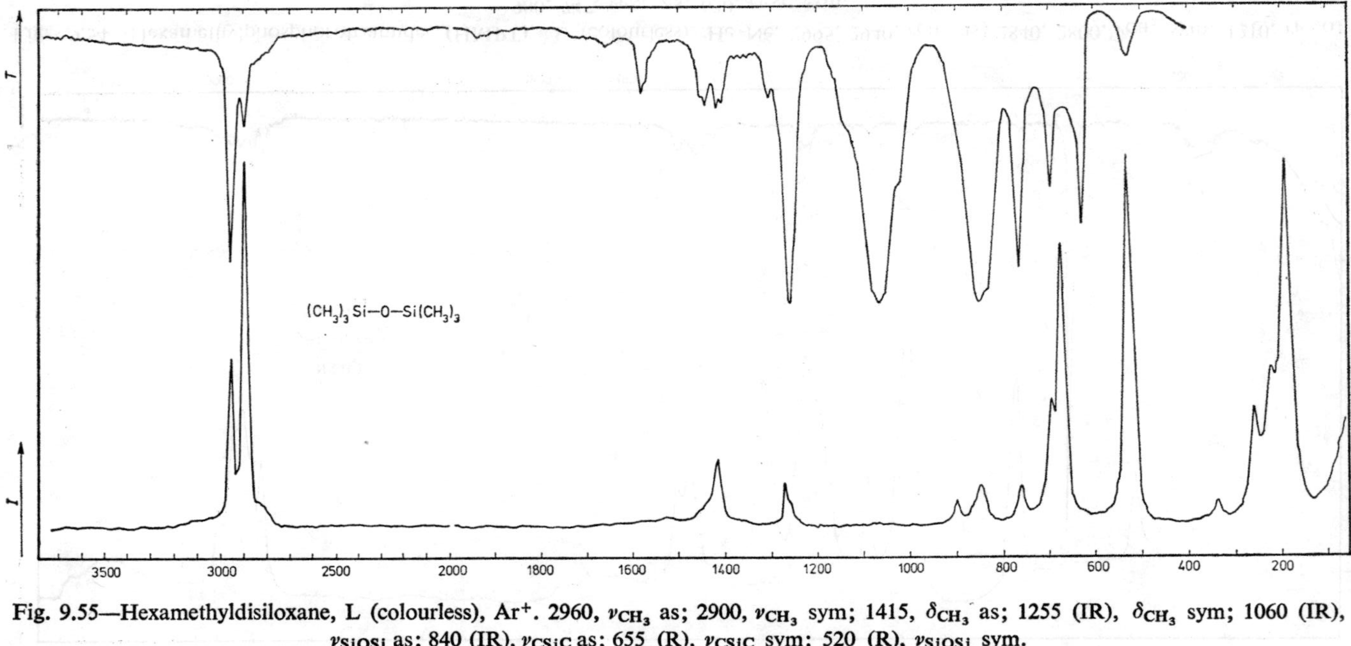

Fig. 9.55—Hexamethyldisiloxane, L (colourless), Ar^+. 2960, ν_{CH_3} as; 2900, ν_{CH_3} sym; 1415, δ_{CH_3} as; 1255 (IR), δ_{CH_3} sym; 1060 (IR), ν_{SiOSi} as; 840 (IR), ν_{CSiC} as; 655 (R), ν_{CSiC} sym; 520 (R), ν_{SiOSi} sym.

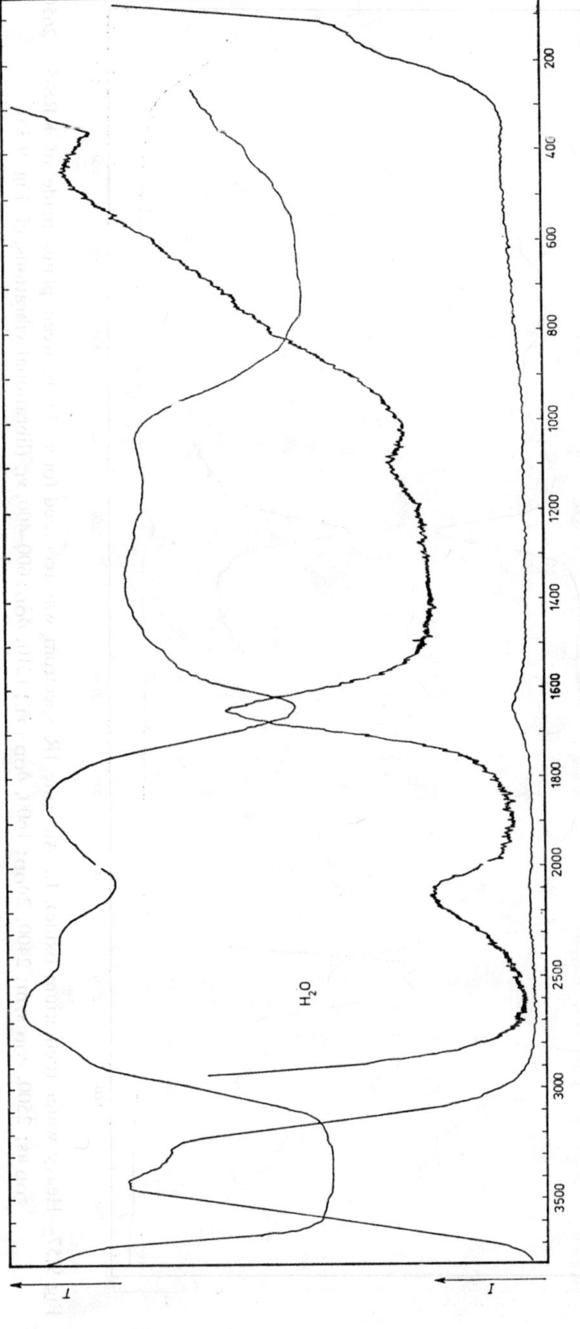

Fig. 9.56—Water, L, Ar^+ (the IR spectrum was recorded for a film between plates made of KRS-5). 3600, ν_{OH} as; 3400, ν_{OH} sym; 3250, $2\delta_{OH}$; 2400, $\delta_{OH}+\nu_L$; 2070, $3\nu_L$; 1640, δ_{OH}; 800–400, ν_L (ν_L—librational vibration, i.e. the vibration consists of an incomplete rotation).

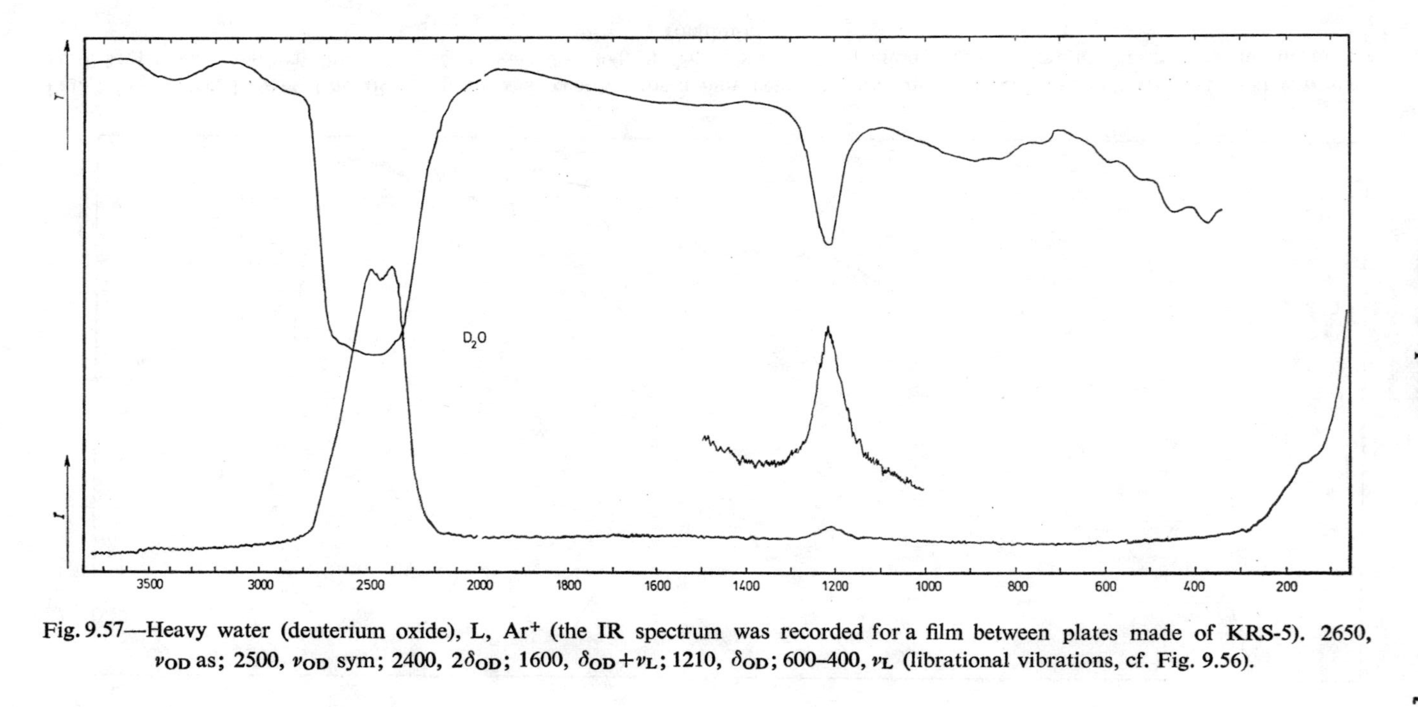

Fig. 9.57—Heavy water (deuterium oxide), L, Ar^+ (the IR spectrum was recorded for a film between plates made of KRS-5). 2650, ν_{OD} as; 2500, ν_{OD} sym; 2400, $2\delta_{OD}$; 1600, $\delta_{OD}+\nu_L$; 1210, δ_{OD}; 600–400, ν_L (librational vibrations, cf. Fig. 9.56).

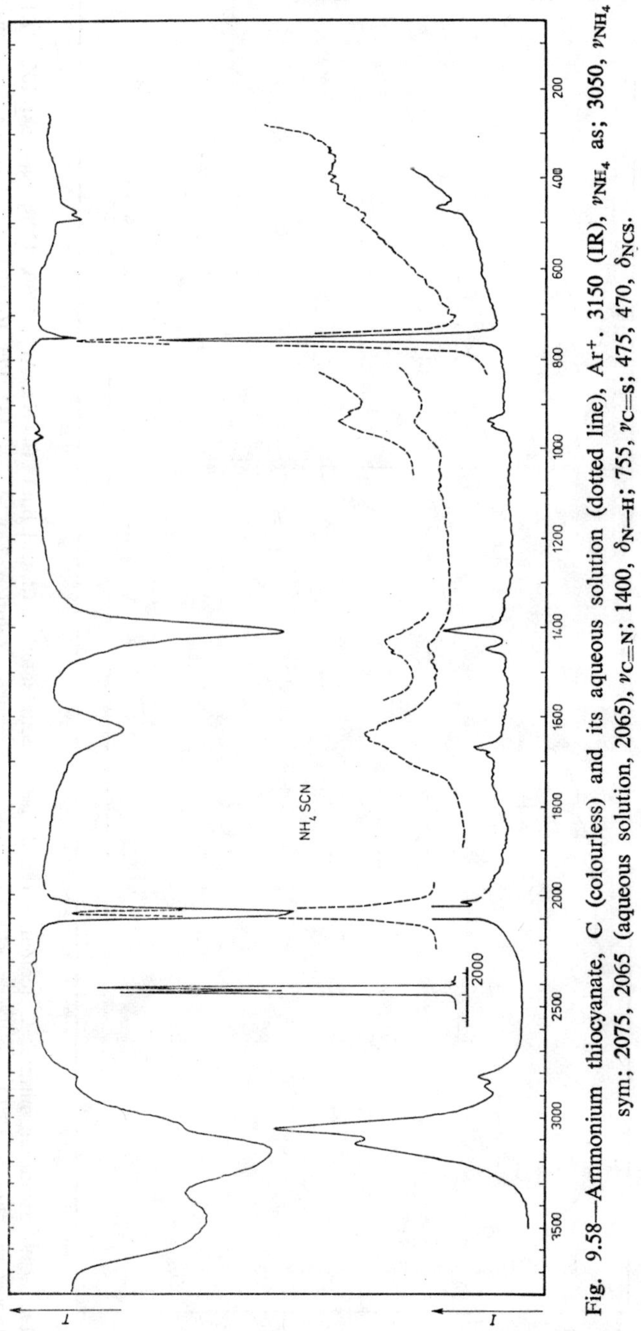

Fig. 9.58—Ammonium thiocyanate, C (colourless) and its aqueous solution (dotted line), Ar^+. 3150 (IR), ν_{NH_4} as; 3050, ν_{NH_4} sym; 2075, 2065 (aqueous solution, 2065), $\nu_{C\equiv N}$; 1400, $\delta_{N—H}$; 755, $\nu_{C=S}$; 475, 470, δ_{NCS}.

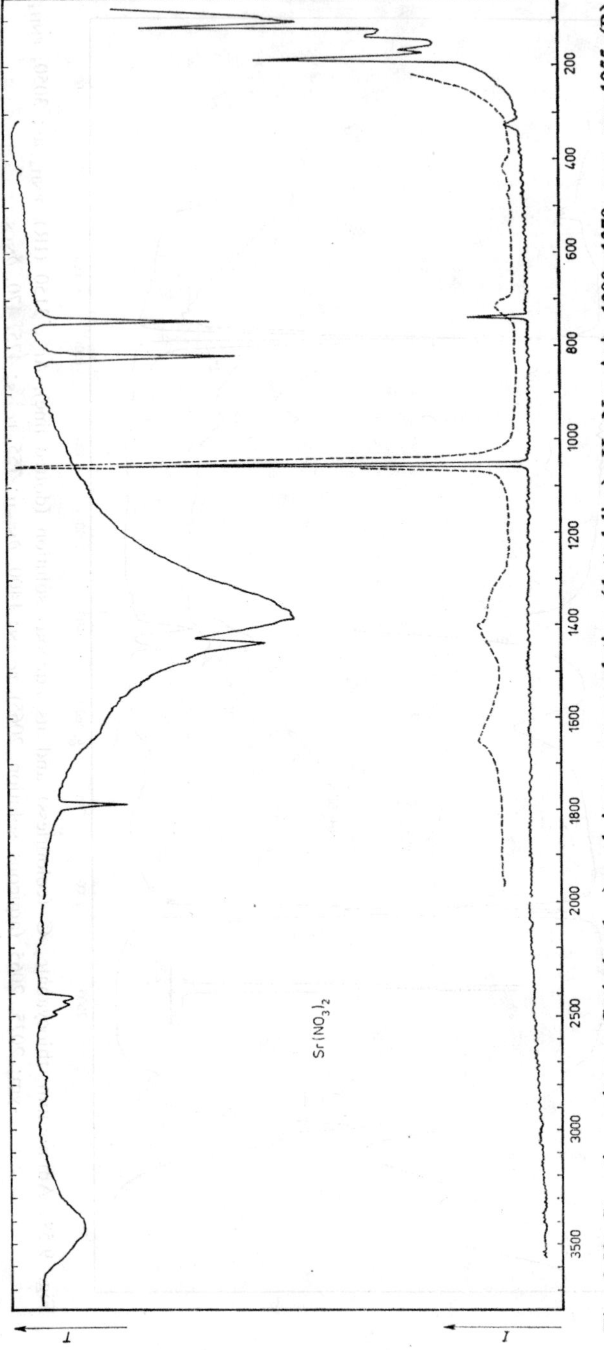

Fig. 9.59—Strontium nitrate, C (colourless) and its aqueous solution (dotted line), He–Ne. **Ar**$^+$. 1390, 1370, ν_{NO_3} as; 1055 (R) (aqueous solution, 1050), ν_{NO_3} sym; 820 (IR), π_{NO_3}; 740 (aqueous solution, 720), δ_{N-O}; 182, 175, 108 (R) crystal latice vibrations.

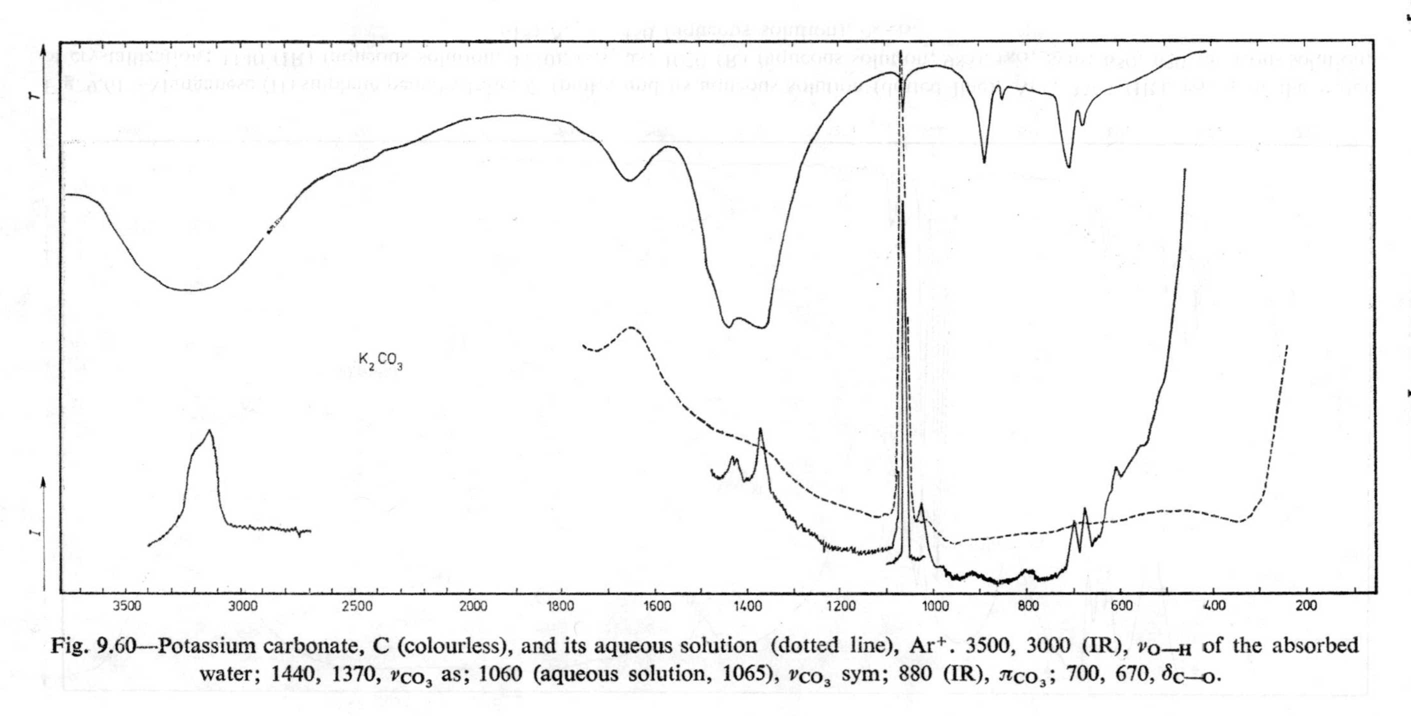

Fig. 9.60—Potassium carbonate, C (colourless), and its aqueous solution (dotted line), Ar^+. 3500, 3000 (IR), $\nu_{O—H}$ of the absorbed water; 1440, 1370, ν_{CO_3} as; 1060 (aqueous solution, 1065), ν_{CO_3} sym; 880 (IR), π_{CO_3}; 700, 670, $\delta_{C—O}$.

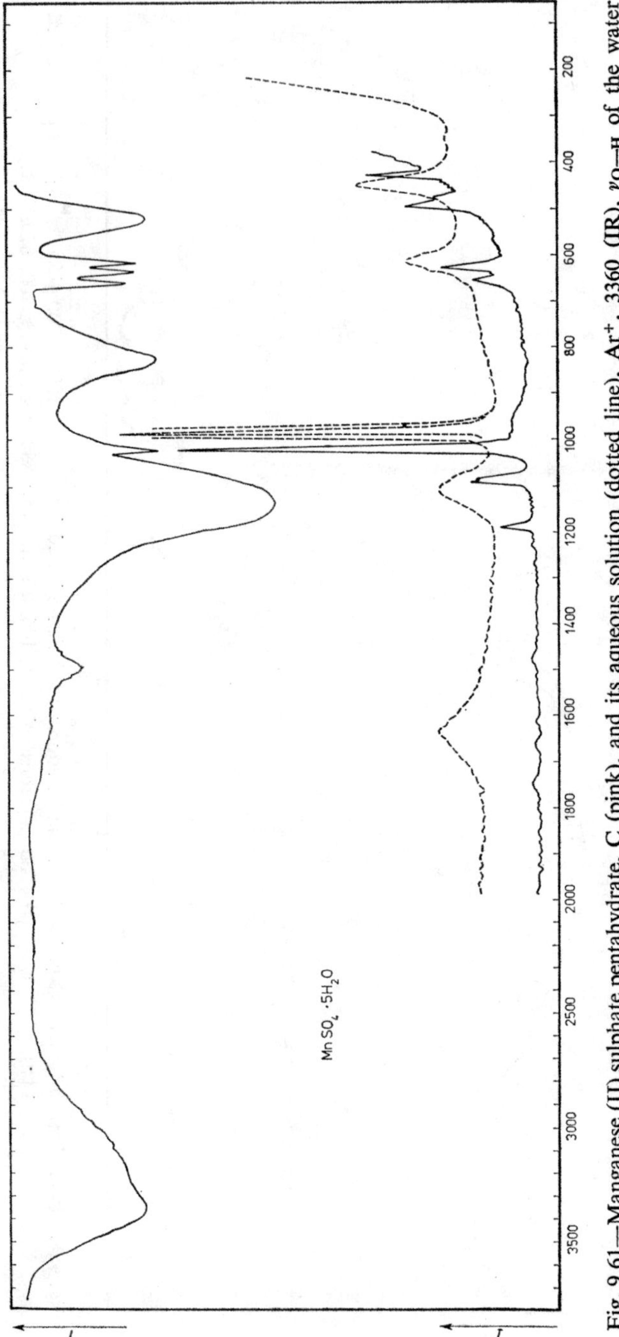

Fig. 9.61—Manganese (II) sulphate pentahydrate, C (pink), and its aqueous solution (dotted line), Ar^+. 3360 (IR), $\nu_{O—H}$ of the water of crystallization; 1140 (IR) (aqueous solution, 1110), ν_{SO_4} as; 1020 (R) (aqueous solution, 985), ν_{SO_4} sym; 650, 620 (aqueous solution, 615) $\delta_{S=O}$; 450 (aqueous solution), $\delta_{S=O}$.

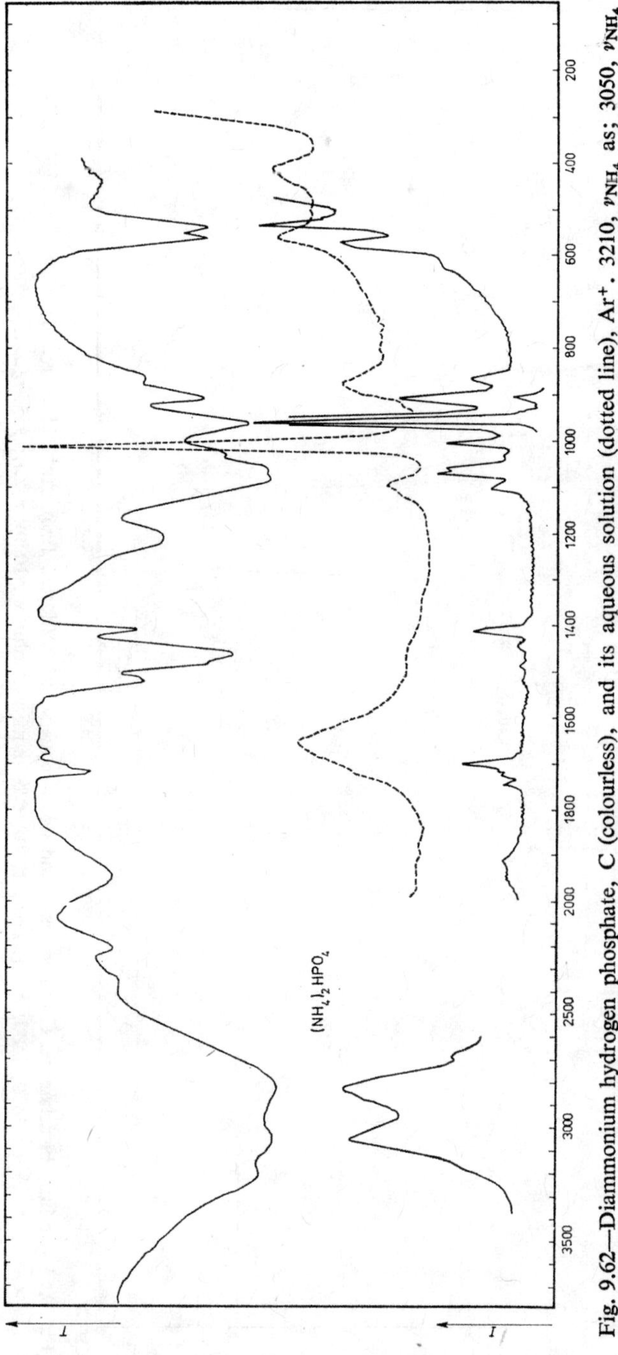

Fig. 9.62—Diammonium hydrogen phosphate, C (colourless), and its aqueous solution (dotted line), Ar^+. 3210, ν_{NH_4} as; 3050, ν_{NH_4} sym; 2815, $\nu_{O—H}$; 1460, $\delta_{O—H}$; 1405, δ_{NH_4}; 1070, 1060 (aqueous solution, 1085), ν_{PO_3} as; 955 (aqueous solution, 990), ν_{PO_3} sym; 865, $\nu_{P—C}$; 550, 525 (aqueous solution, 540), δ_{OPO}.

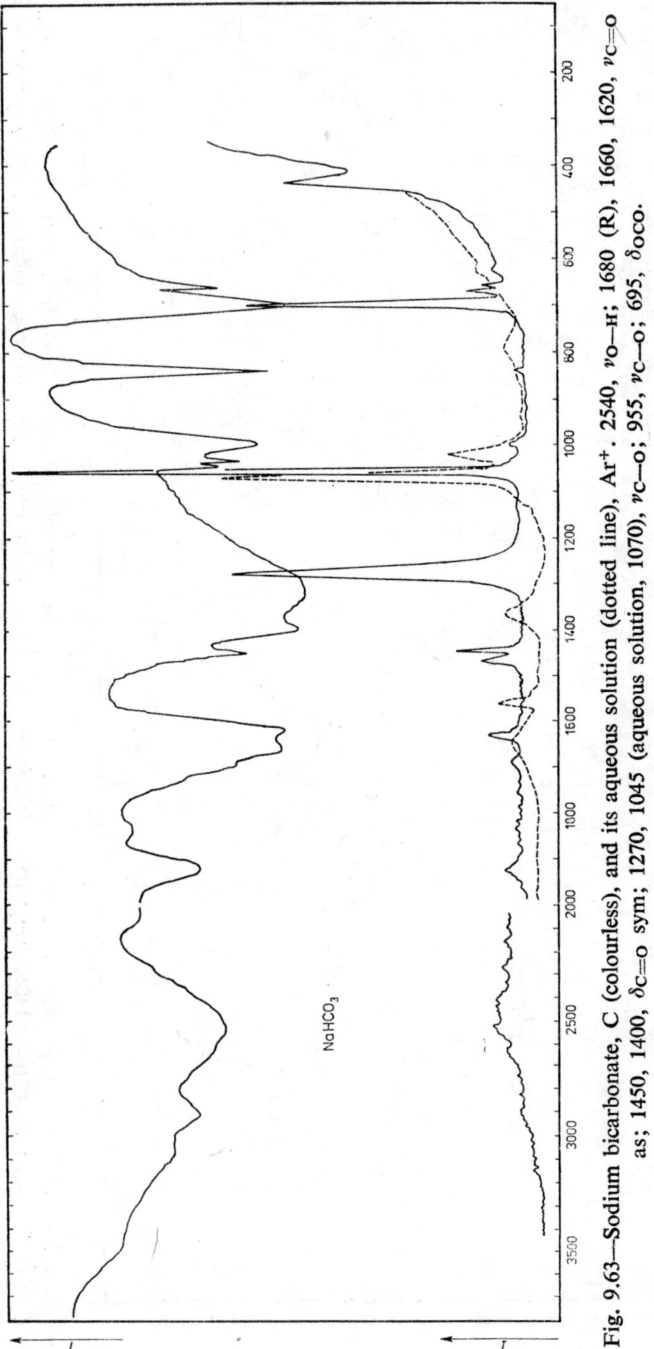

Fig. 9.63—Sodium bicarbonate, C (colourless), and its aqueous solution (dotted line), Ar^+. 2540, $\nu_{O—H}$; 1680 (R), 1660, 1620, $\nu_{C=O}$ as; 1450, 1400, $\delta_{C=O}$ sym; 1270, 1045 (aqueous solution, 1070), $\nu_{C—O}$; 955, $\nu_{C—O}$; 695, δ_{OCO}.

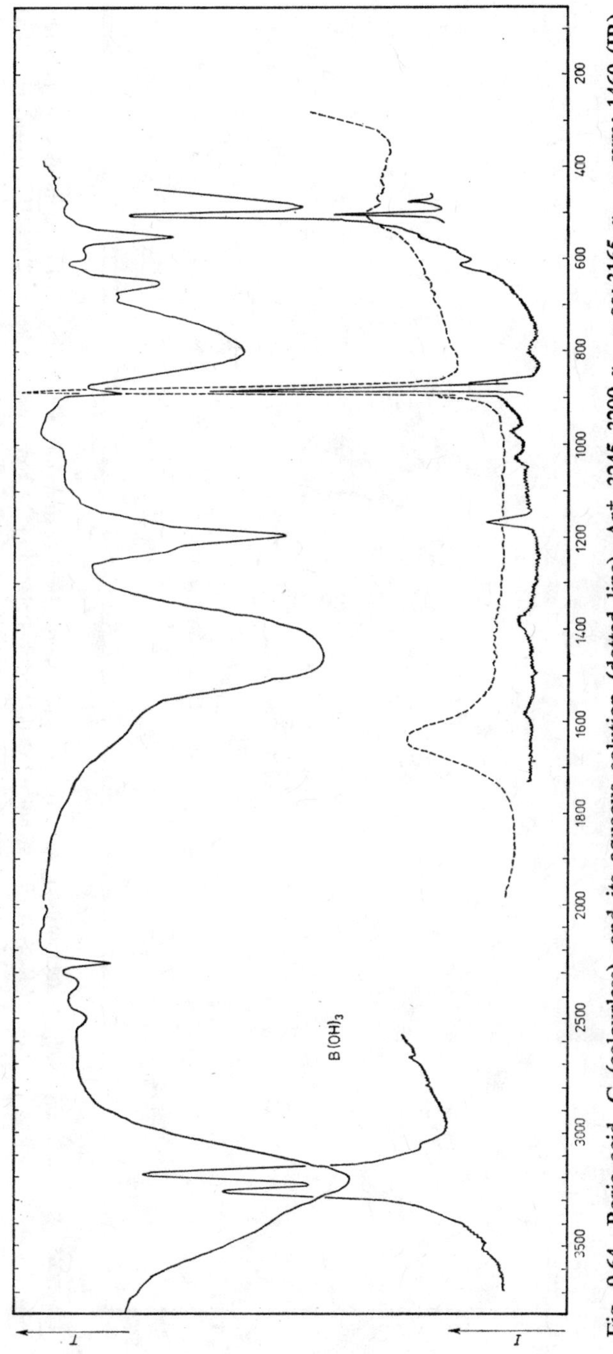

Fig. 9.64—Boric acid, C (colourless), and its aqueous solution (dotted line), Ar^+. 3245, 3200, $\nu_{O—H}$ as; 3165, $\nu_{O—H}$ sym; 1460 (IR), $\nu_{B—O}$ as; 1195, δ_{BOH}; 880 (aqueous solution, 880), $\nu_{B—O}$ sym; 790 (IR), π_{BO_3}; 645 (IR), π_{BOH}; 545, δ_{OBO} as; 500 (R), δ_{OBO} sym.

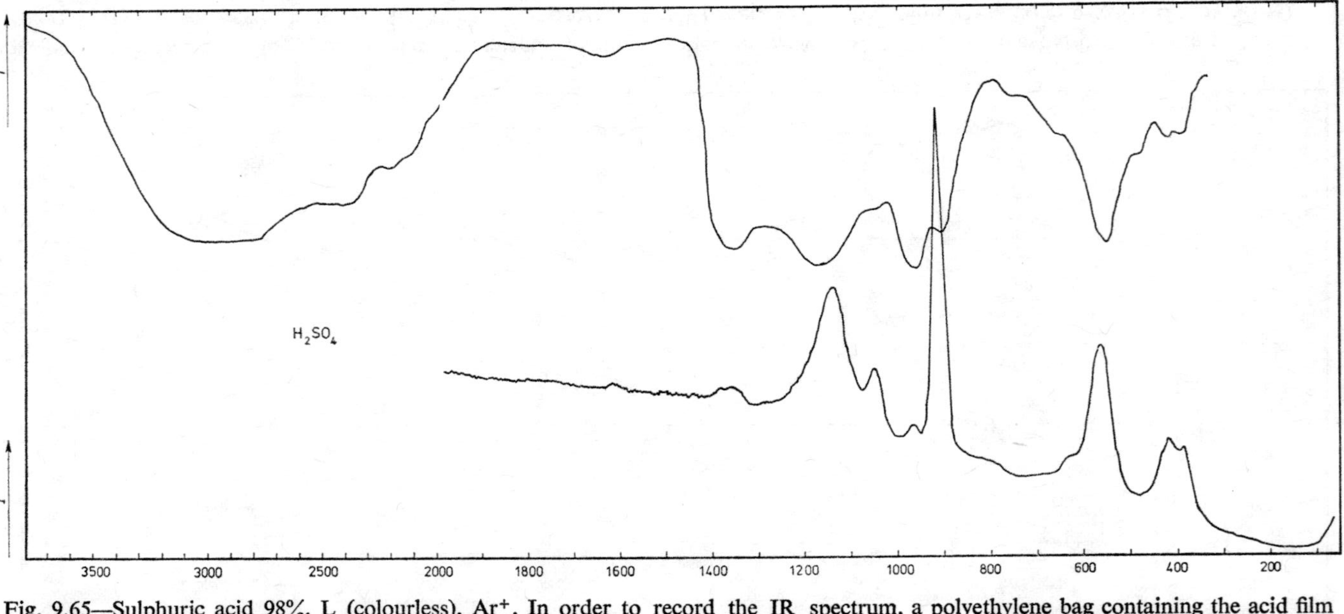

Fig. 9.65—Sulphuric acid 98%, L (colourless), Ar$^+$. In order to record the IR spectrum, a polyethylene bag containing the acid film was placed between plates made of KBr. A similar polyethylene film was placed in the compensating beam. 2900 (IR), ν_{O-H}; 1360 (IR), ν_{SO_2} as; 1140, ν_{SO_2} sym; 960 (IR), ν_{S-O} as; 910 ν_{S-O} sym 560, δ_{SO_2}; 385, δ_{S-O}.

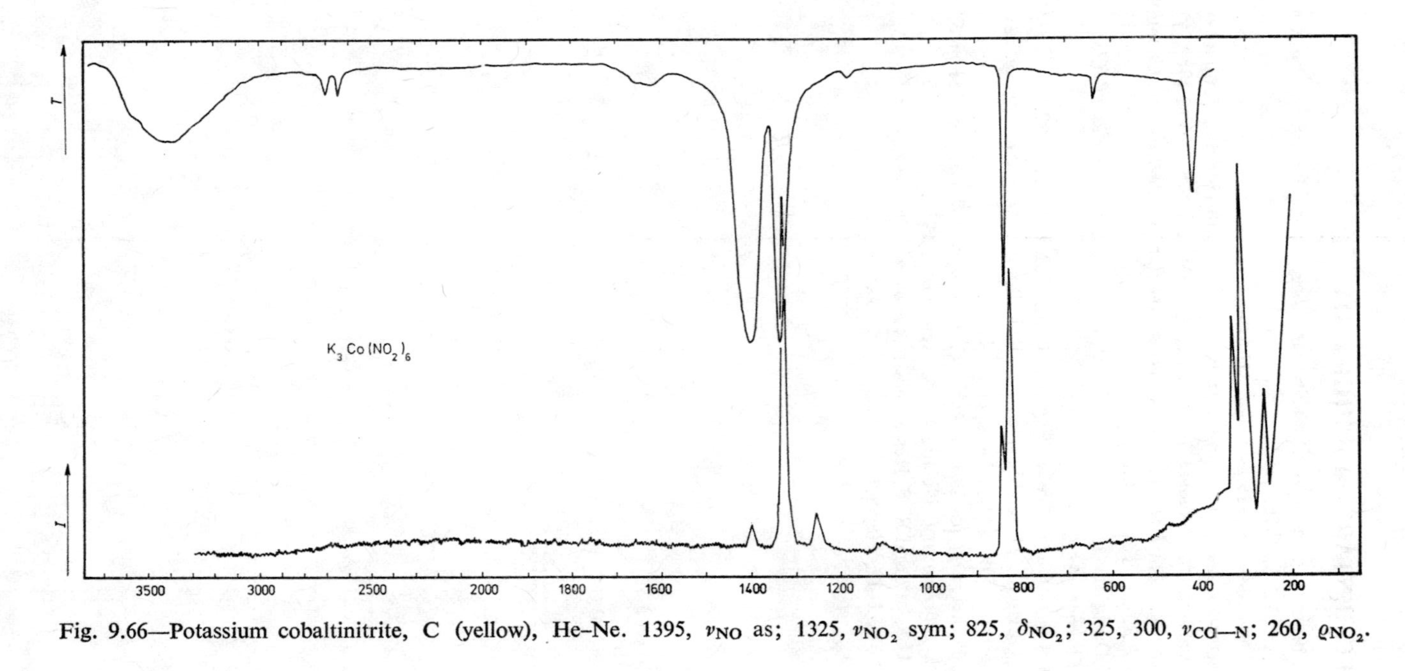

Fig. 9.66—Potassium cobaltinitrite, C (yellow), He–Ne. 1395, ν_{NO} as; 1325, ν_{NO_2} sym; 825, δ_{NO_2}; 325, 300, $\nu_{CO—N}$; 260, ϱ_{NO_2}.

SUPPLEMENTARY BIBLIOGRAPHY

[1] E. G. Brame, Jr. and J. G. Grasselli (eds.), *Infrared and Raman Spectroscopy*, Dekker, New York, 1976.

[2] W. F. Murphy (ed.), *Proceedings VIIth International Conference on Raman Spectroscopy*, Linear and Non-linear Processes, North-Holland, Amsterdam, 1980.

[3] J. Lascombe (ed.), *Raman Spectroscopy VIII*, Wiley, New York, 1982.

[4] *Proceedings IXth International Conference on Raman Spectroscopy*, Chemical Society of Japan, Tokyo, 1984.

[5] J. R. Durig (ed.), *Vibrational Spectra and Structure: A Series of Advances*, Vol. 5, Elsevier, Amsterdam, 1976.

[6] D. A. Long, *Raman Spectroscopy*, McGraw-Hill, London, 1977.

[7] J. G. Grasselli, M. K. Snaveley and B. J. Bulkin, *Chemical Applications of Raman Spectroscopy*, Wiley, New York, 1981.

[8] F. S. Parker, *Applications of Infrared, Raman and Resonance Raman Spectroscopy in Biochemistry*, Plenum Press, New York, 1983.

[9] R. J. H. Clark and R. E. Hester (eds.), *Advances in Infrared and Raman Spectroscopy*, Vols. 1–12, Heyden, London, 1975–1984.

Index

(*cont.*)

T. T. Orlovsky — **Chromatographic Adsorption Analysis**
D. Perez Bendito & M. Silva — **Kinetic Methods in Analytical Chemistry**
B. Ravindranath — **Principles and Practice of Chromotography**
V. Sediveč & J. Flek — **Handbook of Analysis of Organic Solvents**
R. M. Smith — **Derivatization for High Pressure Liquid Chromatography**
R. V. Smith — **Handbook of Biopharmaceutic Analysis**
K. R. Spurny — **Physical and Chemical Characterization of Individual Airborne Particles**
K. Štulík & V. Pacáková — **Electroanalytical Measurements in Flowing Liquids**
O. Shpigun & Yu. A. Zolotov — **Ion Chromatography in Water Analysis**
J. Tölgyessy & E. H. Klehr — **Nuclear Environmental Chemical Analysis**
J. Tölgyessy & M. Kyrš — **Radioanalytical Chemistry**
J. Urbanski, *et al.* — **Handbook of Analysis of Synthetic Polymers and Plastics**
M. Valcárcel & M. D. Luque de Castro — **Flow-Injection Analysis: Principles and Applications**
C. Vandecasteele — **Activation Analysis with Charged Particles**
J. Veselý, D. Weiss & K. Štulík — **Analysis with Ion-Selective Electrodes**
F. Vydra, K. Štulík & E. Juláková — **Electrochemical Stripping Analysis**
N. G. West — **Practical Environment Analysis using X-Ray Fluorescence Spectrometry**
F. K. Zimmermann & R. E. Taylor-Mayer — **Mutagenicity Testing in Environmental Pollution Control**
J. Zupan — **Computer-Supported Spectroscopic Databases**
J. Zýka — **Instrumentation in Analytical Chemistry**